33 Simple Weekend Projects

for the Ham, the Student, and the Experimenter

By Dave Ingram, K4TWJ

CQ Communications, Inc.

Library of Congress Catalog Number 97-67120
ISBN 0-943016-17-7

Editor: Edith Lennon, N2ZRW
Technical Advisor: Lew Ozimek, N2OZ
Layout and Design: Elizabeth Ryan and Edmond Pesonen
Illustrations: Hal Keith
Cover Photo: Larry Mulvehill, WB2ZPI

Published by CQ Communications, Inc.
76 North Broadway
Hicksville, New York 11801 USA

Printed in the United States of America.

Copyright © 1997 CQ Communications, Inc., Hicksville, New York

All rights reserved. No part of this book may be used or reproduced in any form, or stored in a database or retrieval system, without express permission of the publisher except in the case of brief quotes included in articles by the author or in reviews of this book by others. Making copies of editorial or pictorial content in any manner except for an individual's own personal use is prohibited, and is in violation of United States copyright laws. While every precaution has been taken in the preparation of this book, the author and publisher assume no responsibility for errors or omissions. The information contained in this book has been checked and is believed to be entirely reliable. However, no responsibility is assumed for inaccuracies. Neither the author nor the publisher shall be liable to the purchaser or any other person or entity with respect to liability, loss, or damage caused or alleged to have been caused directly or indirectly by this book or the contents contained herein.

TABLE OF CONTENTS

PREFACE v

CHAPTER 1: The Joys of Homebrewing......................1
    The Many Faces of Modern Homebrewing................1
    Helpful Tips for Building Electronic Projects! 3
    Packaging Makes the Big Difference 7

CHAPTER 2: Elmer's Sugarcoated Electronic
    Theory 9
    Voltage and Current 10
    The Five Basic Electronic Components 12
    Resistors...13
    Capacitors...14
    Coils 15
    Transformers 15
    Transistors 17

CHAPTER 3: Projects for the Home Station..............23
    Power Strip with "Big Switch" Simplifies
        Station Wiring...24
    Magic Length Coax Cable Extension Has
        Unlimited Applications.....................................27
    Eliminate RF Feedback with This Quick and
        Easy Fix 30
    Installing a Ground System for Your Station...........32
    Quick-Check Your Transmitted Signal Strength
        with This Neat FSM 35
    Deluxe CW Keyer Includes Transceiver Remote
        Control 39
    Easy Restoration Breathes New Life into Old
        Shortwave Receivers...43
    Rejuvenating Golden-Oldie Transmitters 47
    Simple Concept Gives Full Break-In Operation
        with Older Receiver/Transmitter Combos.........51

CHAPTER 4: HF Antennas **55**
 A VERY Important Note 56
 Antenna Wires and Insulators 57
 Types of Coax Cable 58
 Baluns 60
 An Easy "Armstrong" Method for Erecting an
  Antenna 61
 10-meter Delta Loop 63
 Modified Bobtail Antenna Packs a Big DX Punch 68
 The Carolina Windom: A Hot Multiband Skywire! ... 70
 The "El Toro" Special 72
 The Bruce Array: A High-Powered Antenna! 74
 "Invisible" and "Secret" Antennas 76

CHAPTER 5: Special Treats for VHF Enthusiasts **79**
 Mobile VHF/UHF Antenna and Mount for Trunk
  or Hood Installation 80
 Voltage Spike and Polarity Protection for Mobile
  Transceivers 83
 Heavy-Duty Alternator Noise Filter 84
 Quick and Easy Battery Monitor for
  FM Handhelds 85
 Quick and Easy Mobile Charger for Handhelds 88
 Triple Rate Charger for Home Use 91
 Key Adapter for Working OSCAR with FM
  Handhelds 93
 Turnstile Antennas for Satellite Operations 98

**CHAPTER 6: Big-Time Accessories for Your HF
 Mobile Rig** **101**
 Homebrew a High-Performance Mobile Antenna 102
 Radio-Active Antennas 108
 Phased Antennas: A Serious Mobileer's Delight 110
 Installing a Great Mobile Ground System 112

CHAPTER 7: A Potpourri of Fun Projects **117**
 Blinking Eyes Troll 117
 Electric Wiener Roaster 121
 Soil Dampness Tester for Home Plants 123
 The Micronauts: Two Just-for-Fun Pocket
  Transmitters 125
 Historical Telegraph System Really Works 129
 Build This Classic Transmitter from Amateur
  Radio's Golden Past 134
 Conclusion 142

APPENDIX **A1, A2**
INDEX **I1, I2**

PREFACE

Far too many people believe that building electronic projects at home, what we call homebrewing, is a special skill possessed by only highly experienced and technically inclined radio amateurs, those with an electronics background, or very serious hobbyists. This is not true.

Homebrewing is a very enjoyable, even profitable, pursuit open to everyone regardless of electronic background, training, experience, or license class! The key is starting with simple yet interesting projects that you can actually use, and then progressing to more advanced endeavors as you become more comfortable working with electronics.

Homebrewing various radio station accessories or other items is both fun and rewarding (nothing beats answering questions about a piece of equipment with "Yes, I built it myself!"), but it can be intimidating, especially when projects you may encounter in monthly amateur magazines and other publications can be quite complex. Realizing that, I wrote this book as a hands-on guide to homebrewing for the inexperienced. Every chapter contains a wealth of knowledge, tips, ideas, simple projects, and station enhancements for all amateurs, from beginners to old pros, as well as for anyone else who is interested in building something with his or her "own hands."

The projects in this book include accessories for handheld transceivers and home or mobile stations, antennas, electronic "conversation pieces," and authentic reproductions of old-time communications gear. I also show you how to set up an "ultra-low-cost" HF station that will make you the envy of the airwaves—this group of projects alone is worth the cost of this book. But in addition to the projects themselves, every page is jam-packed with general information and building tips that you'll use for many years to come (perhaps I should have selected the title "The Newcomer's Book of Knowledge"!).

I'd ask the experienced and "seasoned" amateurs who may be reviewing this book to think back to their own first attempts at "dinking" with electronics before pronouncing the contents "too basic." Projects that seem simple to amateurs with 20 years of experience will be new and vital to the newcomer. But no matter what your experience level, I'd suggest checking out the contents: more than a few of the projects may have eluded the most "advanced" homebrewer!

Preparing this book was an extensive undertaking, and I'd like to thank some people for their assistance. Special thanks to Don Stoner, W6TNS, for suggestions and insights on various projects. Thanks, also, to Evelyn Garrison, WS7A, for advice on "attractive projects" (which I included) and "less appealing projects" (which I sidestepped!). A most special thanks to my wife Sandy, WB4OEE, for her invaluable assistance with this manuscript, preparation of drawings, proofreading and "visualizing the projects and their assembly" from a Technician-licensee's viewpoint. All of our efforts will be rewarded if you find this book beneficial in your homebrewing efforts and your amateur radio life.

So welcome to the wonderful world of homebrewing. It is a fascinating area that you can enjoy almost anywhere and anytime, and the fruits of your efforts will raise your self-esteem more than 10db! Now relax in a comfortable chair, fire up your soldering iron, and let's start on the journey into the exciting world of homebrewing. Enjoy!

73, Dave Ingram, K4TWJ

CHAPTER 1
The Joys of Homebrewing

In the early days of radio—and many years before your author was born!—amateurs necessarily assembled their complete station at home, what we call "homebrewing." The driving force behind their endeavors was quite simple: commercially manufactured, or "store-bought," equipment was scarce, while defunct radios and salvaged parts for homebrewing shortwave gear were available for the asking. Circuits were also simple and people had plenty of spare time for building their own rigs. Technology and lifestyles have changed dramatically since then. Most of us now find spare time a very limited commodity. Besides, modern FM handhelds, mobile rigs, and SSB transceivers are much too complex for home-assembly and are readily available from dealers nationwide anyway.

Still, a sincere desire to self-assemble *some* amateur radio-related equipment (or other piece of electronics) continues to burn among amateurs everywhere. Maybe it stems from our proud heritage; maybe it's just prompted by individual needs. An amateur might assemble a simple receiver just for fun or portable use, for example, or might build an antenna or field strength meter because the style or shape desired is not readily available commercially. Whatever the case, anyone who's tried it will tell you that homebrewing is both fun and rewarding.

The Many Faces of Modern Homebrewing

In my opinion, modern "homebrewing" of amateur radio projects falls into three general categories: combining and/or refurbishing commercially manufac-

Figure 1-1. Electronic "blinking eyes" conversation piece I built several years ago. Detailed assembly instructions are in Chapter 7.

tured equipment to produce a custom station; adapting readily available items to personal needs; and building simple projects right "from scratch." Examples of the first category would be "tuning up" inexpensive older gear to make a neat HF station or modifying an FM handheld for CW use on the OSCAR (Orbiting Satellites Carrying Amateur Radio) satellites. An example of the second category would be a minor reworking of an electronic keyer so that it doubles as a wired remote control. Finally, examples of the third category would be building antennas, field strength meters, and even non-radio related projects (this book isn't necessarily just for hams, you know!).

All of these homebrewing areas are covered in these pages and are presented with you, the new amateur or electronics enthusiast who has limited technical expertise, in mind. I "keep it simple" to minimize frustration, to help you expand your horizons at your own pace, and to give you a fun introduction to the world of homebrewing. There's something in here for everyone, and I'm sure you'll find this collection of projects helpful for many years.

Homebrewing is the perfect way to get the particular antenna, battery charger, mobile ignition noise filter, or even that electronic "conversation piece" you want. And it can often be acquired faster by self-assembly than by waiting for delivery via mail order. But most important is the pride of "doing it yourself." And these skills extend beyond amateur radio. For instance, Figure 1-1 shows an electronic toy that I assembled from scratch a few years ago. Needless to say it's quite an attention grabber!

Could you build something similar? Sure: the only requirements are an ability to follow simple instructions, a steady hand, and good soldering techniques. An elementary knowledge of electronic theory is helpful, but not mandatory. A basic knowledge of Ohm's Law is sufficient for understanding, and even troubleshooting, any of the projects in this book. Additionally, the following chapter reviews that theory in simple terms with an emphasis on how to apply it.

So read on with confidence: your homebrewed projects will be successful!

Helpful Tips for Building Electronic Projects!

The best home-constructed projects are those planned, laid out, and assembled along a logical, well-thought-out path. Construction should begin by studying an item's circuit diagram and visualizing the concept behind it. If your project is a super-miniature one, a "third hand" pc holder and magnifier, like that shown in Figure 1-2, will prove invaluable. Pay close attention to areas prone to wiring mistakes (I'll point them out as we go along). Generally speaking, you should assemble items working from left to right, with input and output sections separated as far as possible to reduce feedback. If more than one transformer is to be used in a project, you should usually mount them at right angles to avoid mutual coupling or hum. Audio lines that run through power supply sections, for example, are prone to pick up 60- or 120-Hz hum. If rerouting such wires isn't feasible, enclosing them within shields is recommended.

The most convenient way to assemble electronic projects involves using perforated phenolic board (typically called "perf" boards), available at RadioShack or other electronic parts suppliers. At the outset, place a circuit's components on the board to help you visualize the space needed and to decide on specific wiring routes. Resistors and capacitors can usually be placed flush on the board with their leads inserted through holes. Take a few extra seconds to make sure that color codes align in the same direction. There are two ways of crossing non-connecting wires on perf board: you can move a wire to the top rather than the bottom of the board (above and below, that is), or you can use insulated wire and route one over the other on the same side of the board. The most popular practice is to run long interconnecting wires on the top side of the board and short component-to-component wires underneath. Power leads are usually connected to a bus line out toward the board's sides, while input and output connections are usually situated near the board's front and rear areas. Some components, such as transformers and switches, are mounted off-board and so dictate a certain type of enclosure for a particular project. Whenever possible, try to obtain a case or cabinet that will compliment your finished project.

Now let's discuss some simple steps to assure your success right off the bat. Before actual circuit assembly, take a few minutes to briefly check each component with an ohmmeter. Yes, I understand you're using all new components, but that doesn't guarantee that they're *good*! I vividly remember home-assembling a lavish SSTV monitor some years ago and troubleshooting it for days before discovering a new capacitor was shorted. As an old pro once told me "Do not assume anything. Confirm the facts yourself!" How true!

Resistors should measure within 10 percent of their marked resistance value (just touch meter leads and read values; don't waste time on exact "right-to-the-ohm" measurements). Capacitors larger than .5 mfd can be checked by touching meter leads to each wire while watching for a brief "kick and return to high resistance" action on the meter. This action should occur each time meter leads are swapped between capacitor wires. If a capacitor reads shorted (zero resistance), discard and replace it. Transistors can also be quick-checked with an ohmmeter to ensure that they're not shorted or open internally. To do so, connect one ohmmeter lead to the base of the transistor and one meter lead to the emitter. If the transistor is good, the meter will read either a high or low resistance, depending on the polarity of the meter test leads. Move the meter lead that's connected to the emitter to the transistor's collector. This should again indicate a similar high or low resistance on the meter. Next, interchange the meter leads and repeat the previous steps. Your meter readings should now be reversed if the transistor is good.

This quick-check is usually sufficient to determine whether a transistor is "good" or "bad." It works like this: the ohmmeter's internal battery provides a forward and reverse bias to the transistor junctions, while monitoring results on the meter. The internal battery in most ohmmeters is reversed, that is the negative meter lead is connected to the battery's positive post. If the negative meter lead (battery positive) is connected to the emitter of a PNP transistor and the positive lead (negative of the internal battery) is connected to the transistor's base, the junction is forward biased. This causes the device to conduct, and the meter reads low resistance. Try this simple evaluation technique on a few known-good transistors to become familiar with its application.

Also, coils provide resistance to AC current flow, but will appear as a "dead short" to an ohmmeter (no resistance to DC flow). Again, it just takes a second to touch meter leads to each wire on the coil's ends to make sure the coil isn't open (defective).

Similar quick-checks are applicable to transformers, potentiometers, and other circuit components. Rest assured that the few minutes that you spend doing

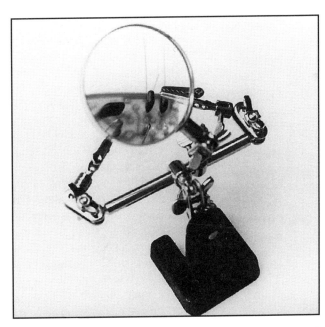

Figure 1-2. Clever "third hand" units that clip-hold circuit boards, connectors, components, subassemblies, etc. for easier assembly and soldering are a real asset in homebrewing. They're available at hamfests and from amateur radio dealers nationwide.

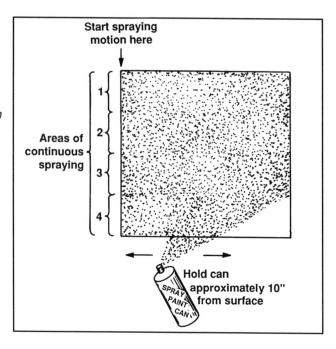

Figure 1-3. Technique for spray-painting a cabinet. Side-to-side strokes with even spray produce optimum results. Discussion in text.

these pretests will be worthwhile in the long run. Knowing that your project has "good components" which will work if wired correctly is always a terrific confidence builder.

Let's now assume that all the components you laid out for assembling a project have been pretested, verified "good," and you're ready to start building on a perf board. Begin by placing each component in its previously selected location, and routing the respective leads to the correct circuit interconnection point(s). Check each component's connection point(s) against the circuit diagram before even considering soldering; then check once more to ensure accuracy. The time spent on this check will repay you tenfold in successful operation the first time around.

Before proceeding to the next component, perform another visual check between the diagram and the circuit you're assembling. This time, be critical of yourself and look for something wrong rather than something right. Hunt for mistakes! Only after you're certain that you can't find an error should you solder the connection and proceed to the next component. Repeat the same procedure on each successive component you add until the assembly is complete. You'll probably have the circuit's diagram memorized by then, but your chance of making and error will be almost zero!

Another technique to ensure that your circuit works the first time power is applied involves checking connections after soldering components. If a coil is wired between a transistor's collector and a common positive voltage line, for example, an ohmmeter check from the transistor's lead, through the transistor socket, and on through the coil winding, the common positive voltage line and back to the ohmmeter ensures good solder connections as well as transistor socket stability. Practice this technique and you'll see how easily the ohmmeter can be used to guarantee correct wiring.

If your completed project doesn't work when powered up, the positive battery lead can be disconnected and routed through the ampmeter to determine if the circuit is drawing current. Each section of the circuit can then be independently connected to the power source in sequence while you monitor its current. This check will identify a faulty section and direct you to the exact problem area. Another technique, known as dynamic troubleshooting, often proves beneficial as well. Apply an appropriate signal to the circuit's input and, beginning at the output and working backwards toward the input, check for a properly processed signal at the output of each stage. When the problem is narrowed to a single stage, switch the input signal source to the defective section to evaluate performance within that section. Continue moving down the signal's path until the defective component is located.

My final suggestion is for you to consider asking a technically knowledgeable friend to look over your project. There's no reason to feel embarrassed over such requests. Indeed, it shows your respect for that person's knowledge. Sometimes even the best builders look at a circuit until they can't see an obvious problem. This is natural and, take it from the voice of experience, it happens more often than we like to admit! Just remember you're new at this—you're not expected to be a pro!

Packaging Makes the Big Difference

Any homebrew project worth building is worth enclosing in a neat cabinet. "Cosmetic appeal" plays an important role in a project's life and in your sense of pride in a job well done. Attractive enclosures are advertised in monthly amateur radio or other hobby magazines and are available from dealers nationwide.

In many cases, cabinet enclosures to match your existing equipment can be purchased separately from a manufacturer. Any homebrewed piece of equipment can be easily assembled in a case which will match your other station equipment. You can also contact a local sheet metal shop to manufacture front panels or to bend and pierce chassis as necessary; just make a heavy paper template with cutouts for knobs, meters, etc. for them to follow. The final cost of such custom work can be surprisingly low, but will result in an item you'll enjoy and be proud of for many years.

A good coat of paint that blends with other station equipment is the final professional touch. A large variety of paint colors are available from all sorts of suppliers, including auto parts stores. Remember to keep some extra paint for any future "touch ups" that might be necessary. Painting is a special technique in itself, but it can usually be mastered in just a few minutes time. The area in which you'll be painting should be between 60 and 90 degrees Fahrenheit, very dry, and well ventilated. Avoid painting during cold or very humid weather as these conditions usually cause the paint to "run." Remember, too, several light coats produce a much better appearance than one heavy coat. When spraying a piece of equipment, position the spray can at least 10 inches from the surface. Spray in full sweeping motions, crossing the area with each motion (see Figure 1-3). In other words, begin spraying before the sweeping motion reaches the project's edge and continue until passing beyond its far edge. Never slow your motion while paint is being sprayed. Allow 30 minutes to one hour drying time between coats, or as directed. About three coats is usually sufficient.

If you'd like a slightly wrinkled look, rather than a smooth surface, here are some suggestions. After the final coat of paint has been applied, move the spray can a slight distance back and spray paint upward. The paint will rise and then fall back onto the surface creating a "leather look." This technique requires a bit of practice, so spend some time painting a scrap item before working with your actual project. Finally, remember to add decals to the front and apply a coat of clear plastic spray for protection.

You can apply these "cosmetic" techniques to your existing station, to the projects in this book, or to any other project you choose to take on. There's really no limit to the applications you'll find for your new creative ideas and skills!

CHAPTER 2
Elmer's Sugarcoated Electronic Theory

All radio amateurs had to study basic electronic theory for their ham licenses, and many non-hams also encountered it in school or on the job. Still, many people naturally experience difficulty applying that knowledge to actual circuits and equipment. Trying to understand electronics is unique in this respect because, unlike mechanical devices, electron flow can't be "seen" in action: it must be detected, measured and/or converted by sensors, such as speakers, meters, test instruments, etc. Once the technical facts are put into perspective and accepted as true working concepts, however, bridging the gap between theory and operation is surprisingly simple. Homebrewing circuits and "dinking" with equipment then becomes an enjoyable pursuit, just as "souping up" autos and building hot rods is to many people.

Let me expand on the automobile analogy a bit here. Some people never raise the hood of their car or know a carburetor from a manifold while others are familiar with nearly every device related to an engine, yet both groups enjoy motoring successfully. But, once equipped with even basic knowledge, any motorist will feel more confident when driving, just as an amateur with basic electronics knowledge will feel more confident when using fancy gear.

For this reason, this chapter presents, in plain language, a super-streamlined discussion of electronic theory. I trust you'll find this information enlightening, refreshingly different from "hard nose theory," and beneficial for working with all types of amateur radio equipment and projects. In other words, this chapter is your "confidence builder," written to help you enjoy both ham radio and homebrewing.

Simply stated, electronics involves two characteris-

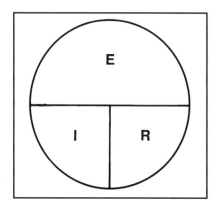

Figure 2-1. Basic "circle formula" as applied to Ohm's Law. This same "circle" concept works for all other formulas. Discussion in text.

tics, two forms of electron flow, and only five types of components. Electron flow can be used to "make things move," (like "receive" signals that are invisible voltages moving in the air), to convert them to frequencies we can hear, and then to boost them in strength so that they can actuate a speaker cone which produces the sounds detectable by our ears. Electron flow can also be used to measure the level of such signals and to position the needle of a meter to indicate the strength of the flow.

Every electronic device requires the flow of electrons to operate. Let's take TVs as an example. In these devices, only one small stream of electrons traveling in a beam makes the pictures on the screen. Have you ever noticed that single dot in the center of an older TV immediately after it's turned off? When the set is operating, that single dot races back and forth across the screen, making 480 horizontal lines, 30 times a second, and the screen's phosphor coating retains all the lines to create the picture you see.

Voltage and Current

As I mentioned above, electron flow has two specific characteristics: voltage and current. We can compare these characteristics to water flowing in a garden hose. The amount of water is analogous to current, and the pressure pushing that water is analogous to voltage. Electrons move from the negative terminal to the positive terminal in a circuit, so, likewise, the inlet source (the faucet) of water is comparable to the negative terminal of a battery or power supply, and the outlet of the water (the nozzle) is comparable to the positive terminal. Think about this concept for a minute, then answer this: Will a large or small hose pass the greatest amount of water, and why? The larger hose passes the most water because it has the largest area for flow, or the least *resistance* to flow. The same is true for electronics: the lower the resistance in a circuit, the greater the flow of current—it, too, takes the path of least resistance!

Easy, right? Well, so far we've only discussed current moving in one direction, or direct current (DC). Now let's consider alternating current (AC). Rather than moving in only one direction, AC, as its name indicates, moves back and forth (oscillates) many times per second. The number of alternations, or cycles per second, is known as frequency and is measured in Hertz (Hz), in honor of Heinrich Hertz. (By the way, the designation of Hertz has been in common usage only since about the mid 1960s, so you'll notice radios prior to that were frequency-marked in "kilocycles or megacycles," rather than "kiloHertz or megaHertz." This is just a bit of foreshadowing—we'll soon be talking about it more in "refurbishing old-time radios"!)

Returning to our hose analogy, you might ask what purpose alternating the water flow would serve. Water in a hose, even if it could be made to alternate back and forth, would have no effect on another hose or pipe placed next to it. Herein lies a significant difference between mechanics (water flow) and electronics (current flow): current flowing in a wire produces a surrounding *magnetic field*. If the current "alternates," or "oscillates," the magnetic field will expand and collapse at the frequency of that oscillation (maximum current flow equals maximum field and vice versa). Any wire lying in this moving field will have an electrical current induced in it. The current, in turn, produces its own magnetic field. Winding wire into a coil (or inductor) intensifies this phenomenon. If a current and voltage is induced from one coil (primary) to a nearby coil (secondary), it is called transformer action. The amount of current or voltage induced in one coil by the other may be stepped up or down based on the relative number of turns in each of the two coils.

Still using our analogy of electron flow and water, let's discuss amplifying devices like NPN transistors and vacuum tubes. The transistor's emitter or the vacuum tube's cathode is comparable to the inlet on a water faucet, the transistor's collector or vacuum tube's plate is comparable to the faucet's outlet, and the transistor's base or vacuum tube's grid is comparable to the

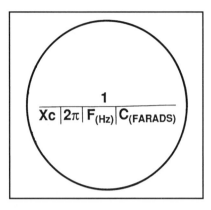

Figure 2-2. Formula for capacitive reactance utilizing "circle formula concept." Hold your finger over the unknown quantity and perform function indicated to determine capacitive reactance, frequency, or capacitance.

faucet's handle. Rotation of the faucet handle varies the amount of water which flows in the pipe (output). Assuming you could move the faucet handle back and forth at an extremely fast rate, you would effectively "modulate" the output of water at that same rate. Just a small rotational force on the faucet handle, however, produces a significant change in output force, thus amplification (of force) has taken place. Now assume the handle is preset to a point of 50 percent water flow through the faucet (DC current flow), and you rapidly move the handle back and forth (equivalent to an AC modulating signal). The output from the valve reproduces the movement of the valve handle, and represents an AC signal superimposed on DC. This analogy illustrates precisely how transistors and vacuum tubes operate—in fact, the British even refer to vacuum tubes as electron valves!

The Five Basic Electronic Components

When looking into a modern SSB transceiver, there's a natural tendency to marvel at all the complex circuitry. If we look closer, however, we realize it's comprised of only five basic components: resistors, capacitors, coils, transformers, and transistors. These five devices are used in combination to make various stages, the stages are combined to produce various sections, and the sections produce operational entities, such as the receiver, transmitter, and VFO. If you understand the general concept behind each component's function, and you consider only one stage at a time, visualizing how electronic circuits work becomes quite simple.

The following sections describe how each of the "big five" components work and how they're combined to produce additional families of components.

Resistors

The simplest and most common components in electronics are resistors. Simply explained, these devices oppose DC current flow by requiring it to take a path which offers resistance. Carbon is the most common material used for this purpose and is encapsulated in a resistor much like carbon lead is in an ordinary pencil. The only significant difference is that carbon in a resistor is highly compressed and the amount used is carefully controlled to produce a specific resistance, within a certain "tolerance" (the "value" of the resistor).

Would you like to make a resistor right now to help you visualize its "innards"? Rub a heavy trace of lead from a pencil onto some paper. Go back and forth over the line many times to deposit plenty of carbo. Now, switch your trusty ohmmeter to its highest resistance range and touch each of its leads to each end of your pencil mark. Congratulations: you just homebrewed a 2 or 5 megohm resistor! While resistors can be made from other material, including wire, they all function basically the same way—opposing current flow.

Applications for resistors in electronic circuits are quite diverse and vary from dropping a voltage or signal level to serving as a "load" for a transistor or tube's output. Think back to our water faucet analogy: the movement of the handle was reproduced at the output of the water hose, but the change could be detected only if there was a device or load which permitted the output to be seen or measured. By using a resistor as a load, we can measure and use output signal variations caused by variations in a transistor input signal.

Let's now apply Ohm's Law to a typical application requiring a resistor. Remember Ohm's Law? Stated in simple terms, this law says the current flowing into a circuit (I) is equal to the applied voltage (E) divided by that circuit's resistance (R). Mathematically this is stated as $I=E/R$, $E=I/R$, or $R=E/I$. These three laws can be combined into one simple form using a "circle formula," as illustrated in Figure 2-1. Place your finger over the unknown quantity to be calculated, then read the indicated function from the other two parameters.

Let's assume you're gearing up for OSCAR satellite

13

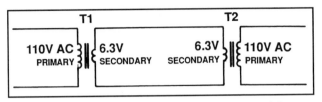

Figure 2-3. Concept of transformer action. The voltage applied to primary causes expanding and collapsing magnetic field that induces voltage into secondary. Amount of the voltage induced depends on turn ratios of primary and secondary windings.

operation and managed to find a used 2-meter transverter at a hamfest. The vacuum tube-type transverter lacks an internal power supply and requires 250 volts DC at 50 milliamps (mA) plus 12 volts AC for operation. Let's further assume your (vacuum tube-type) HF transceiver has an accessory socket on the rear providing 300 volts at 75 mA plus 12 volts AC (for filament power). The transverter can be powered from the transceiver, as its demands will not exceed the transceiver's accessory output capabilities. A dropping resistor can be used to reduce the transceiver-provided 300 volts to 250 volts. This step is easily accomplished by inserting a series resistor to drop 50 volts while passing 50 mA of current. According to Ohm's Law, this calculates as follows: 50 mA = .05 amperes, so 50 volts ÷ .05 ampere = 1000 ohms. Now look again at the circle formula in Figure 2-1 and cover the R. The E appears over the I, which means you divide the top (E) by the bottom (I) to find the unknown (R). The power dissipated by this resistor is calculated as: 50 volts x .05 amperes = 2.5 watts; thus a 5-watt resistor can be used. The HF transceiver was capable of supplying up to 75 mA of current. Since the 2-meter transverter required only 50 mA, or .05 ampere, the power source is not overloaded.

Capacitors

The basic function of a capacitor is to pass AC while blocking (and storing charges of) DC. Despite readily passing AC, capacitors also provide a resistance to AC. This AC resistance is identified as capacitive reactance (X_c) and may be calculated by the formula: $X_c = 1 \div 2\pi \times F$ (Hz) x C (Farad) (also see Figure 2-2). Assuming a 60-Hz signal is applied to a 1-microfarad (mfd) capacitor, its reactance will be: $1 \div 6.28 \times 60$ (Hz) x 1 x 10^{-6} (1 mfd) = 2650 ohms. If a 1000-Hz signal is applied to that same 1-mfd capacitor, the reactance will decrease as follows: 1 ÷ 6.28 x (1 x 1000) (Hz) x 1 x 10-6 (1 mfd) = 159 ohms. A coupling capacitor of 1 mfd may thus be used to block low frequency (60 Hz) signals while simultaneously passing high frequency (1000 Hz) signals.

In the same way resistors are used to drop DC voltages, capacitors can be used to drop RF voltages. One example is variable (tuning) capacitors often located in the input section of VHF amplifier circuits. A variable capacitor connected between the input and base of a transistor varies input drive, while another variable capacitor between the transistor's base and ground completes the voltage divider network and matches reactance or impedance. If this explanation seems slightly complex, just visualize variable capacitors as blocking DC while acting like variable resistors to AC (signals).

Coils

A coil of wire is used to pass DC while opposing (but not completely blocking) AC. Its function goes slightly further, however, as the constantly changing magnetic field resulting from the coil's opposition to AC current is used to induce a voltage or current into an adjacent coil. This effect is utilized in transformers, which we'll discuss presently.

Stand-alone coils also have some interesting applications. One such application is as an RF choke inserted between a DC power supply lead and an RF-amplifying transistor's collector. In this case, the RF choke (coil) passes DC to power the device (and circuit), while blocking RF energy to signal-isolate the stage. Generally speaking, small coils are used in high frequency applications (RF frequencies), while large coils are used in low frequency applications like power supplies (60 or 120 Hz).

Transformers

Transformers are similar to the previously-discussed coils in that they are classified for use at audio frequencies and at radio frequencies. They differ in that they're not used with DC (they then would become a large electromagnet!). The newcomer to electronics will normally visualize a transformer as a device which applies an incoming voltage or signal to a

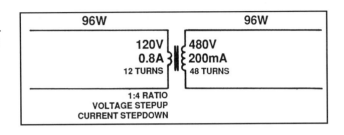

Figure 2-4. Example of step-up ratio of 1 to 4 in transformer.

primary winding and extracts an output from the secondary winding. A transformer actually works in either direction, as illustrated in Figure 2-3. Referring to the figure, you see that 110 volts AC is applied to the primary of T1, and 6.3 volts AC is generated in the secondary winding. If the secondary of T1 is connected to the secondary winding of T2, the primary winding of T2 will then have 110 volts AC induced in it. Looking at the previous figure, you might reasonably ask why you should use two transformers rather than simply applying the 110-volt AC source directly. The answer is that such an arrangement provides isolation. One side or wire of a home 110-volt AC line is "hot," and the other side or wire is "ground." This presents a potential shock hazard. By using two transformers connected back to back, both sides of the AC line have been isolated or lifted above ground to produce an extra margin of safety. Looking closer at the previous example, let's assume an experimenter needs approximately 60 rather than 120 volts. A large dropping resistor could be used to reduce the voltage (assuming load current is known); however, a more desirable solution is to use the center tap connection of T1 to deliver half-voltage AC to T2.

One of the basic laws of physics is that energy cannot be created or destroyed, but only changed in form. We see this law demonstrated in transformers. As an example, notice that the transformer shown in Figure 2-4 has a step-up ratio of 1 to 4. This means that there are four times as many turns in the secondary winding as in the primary winding. If 120-volts AC is applied to the primary, approximately 480 volts will be available at the secondary winding's output. Current steps down when voltage steps up, thus approximately 200 mA of current can be drawn from the secondary. This is because 120 volts x .8 ampere = 96 watts (primary) and 480 volts x 200 milliamperes = 96 watts (secondary). The previous illustration confirms that transformers only convert one form of energy to another. To be absolutely correct, however, I must point out that in actual operation only 65 to 80 watts will be available on the transformer's primary. The difference between input and output power is the amount dissipated by the transformer in the form of heat. This is why many transformers operate "warm to the touch." A slight loss is always encountered when transforming energy.

Transformers operate on the principle of a changing magnetic field produced by the applied AC voltage, thus they also produce hum that can be induced into other transformers mounted in close proximity. This mutual coupling effect can be minimized by mounting transformers at right angles to each other, in the manner shown in Figure 2-5. Remember that simple-yet-effective tip when home-constructing power supplies. Generally speaking, transformers fall into two categories: AF (or power transformers) and RF transformers. AF and power transformers operate at low frequencies and employ iron cores, while RF transformers operate at very high frequencies and only have an air core.

Transistors

Concluding our discussion of the electronics "big five" components is the transistor. Stated in the simplest of terms, the fundamental property of a transistor is that a current in the base-emitter junction causes a larger current to flow in the collector-emitter junction. The flow in the collector-emitter junction will be controlled by the amount of current in the base-emitter junction. Thus a small varying signal applied to a transistor base varies the current flowing (from negative to positive) between its collector and emitter, even though the output (collector-emitter flow) has been amplified by the transistor. As illustrated in Figure 2-6, the transistor output appears across the transistor's related load resistor (R3), and since only the variations in that DC current are recognized as an AC signal by the coupling capacitor (C2), only these variations are transferred to the next stage. This same coupling capacitor also isolates the DC collector voltage of the stage illustrated from the base of the following transistor. (See earlier discussion on capacitive reactance for more details.)

There are two basic types of transistors: PNP and NPN. The NPN transistor (illustrated in Figure 2-6) is

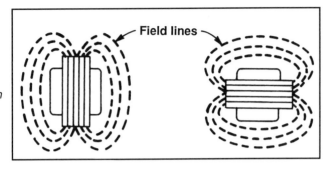

Figure 2-5. Example of how transformers can be mounted at right angles to minimize mutual coupling and "hum effects."

very common in electronic circuits and popular among radio amateurs because its operation is similar to a vacuum tube with respect to polarities. The emitter is connected to a power supply or battery's negative terminal while the collector is connected (through the load resistor) to the power supply or battery's positive terminal. As an incoming signal causes base voltage fluctuations, the collector current changes accordingly. Adding more negative bias to the base of the transistor (by decreasing the base-to-emitter resistor's value) biases the transistor closer to cutoff and is called reverse biasing. Adding more positive bias to the transistor's base (by decreasing the base to collector resistor's value) biases the transistor closer to saturation (maximum current flow) and is called forward biasing. In the majority of transistor circuits, the previously mentioned transistor base resistance establishes a transistor operating point. Varying the incoming signal applied to the base instantly forward-and-reverse biases the transistor around that reference point. The ratio of forward and reverse bias resistors also determines the transistor's function. If, for example, the base-to-positive voltage resistance in Figure 2-6 is 500k ohm (R1) and the base-to-negative voltage resistance is 1k ohm (R2), the transistor is biased at cutoff (reverse bias) until an incoming signal forces it into conduction by forward biasing. Such a circuit might be used in a class C amplifier.

A few sentences ago I mentioned that amateurs favor NPN transistors because their circuit polarities are similar to vacuum tubes. Let's clarify that point before proceeding onto PNP transistors. A vacuum tube produces amplification by varying current between its (negative) cathode and (positive) plate according to a grid-applied signal. Although a vacuum tube has higher operating voltages and impedances, a tube, such as a triode, performs the same function as an NPN transistor. I should point out, however, that the elements in a vacuum tube are not physically connected and depend strictly on electron flow between closely spaced elements in a vacuum. Additionally, tubes require only a single grid-biasing resistor to maintain control of plate current. This is because the filament is heated to the point of "boiling off electrons" like a heater in a house. They travel through a grid or screen mesh to the plate, which is connected to the positive battery lead and accelerates their journey. Without grid bias "holding back" the flow of electrons, the plate will saturate. Reverse bias, so to speak, is therefore all that's necessary to control this electron flow. In the same way that heat radiates from a hot element (filament), current travels only one way (from filament/cathode to plate) in a vacuum tube. But don't assume that seemingly "old-fashioned" vacuum tubes are of no interest today. Quite simply, they're the most effective and economical means of obtaining ultra-high RF power levels. Commercial transmitters running upward of 10000 watts necessarily utilize very large vacuum tubes to achieve such high power levels. In the case of legal-limit amateur radio equipment, large transmitting tubes, like the popular 3-500Z, continue to offer a cost-efficient advantage.

A PNP transistor functions identically to an NPN transistor, except that all operating voltages are reversed. The PNP's collector potential will thus be negative while its emitter potential will be positive. When positive bias is applied to the base, the transistor is reverse biased and less collector current flows.

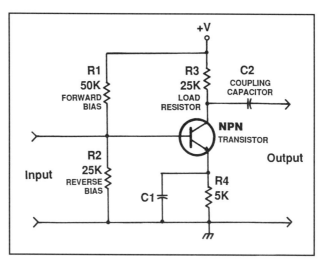

Figure 2-6. Example of electronic reasoning and current flow...in other words, how the critter works!

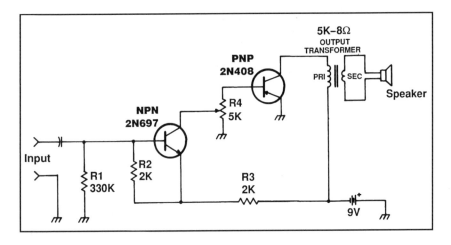

Figure 2-7. Combining NPN and PNP transistors in a two-stage audio amplifier using one battery has several benefits. As the volume is lowered, the current drain on the battery also decreases.

Likewise, when the base is forward biased with a negative voltage, more collector current flows. In other words, the PNP transistor is a "reverse copy" of the NPN transistor. Each type of transistor serves a particular purpose, and different types may even be used in combination to achieve interesting results.

One example of this is illustrated in Figure 2-7. This circuit shows a simple yet effective audio amplifier, similar to the type used in radios and FM handhelds. Notice that only one battery is used and both PNP and NPN transistors are employed. There are two resistors on the NPN transistor's base: one (R1) connects to the battery positive (through ground) and the other (R2) connects to the battery negative through the 2-k ohm resistor (R3). These resistors establish a proper no-signal biasing point for the NPN transistor. The PNP transistor also has two bias resistors connected to its base: a 5-k ohm potentiometer (R4) and the (internal) collector-to-emitter resistance of the NPN transistor. As an incoming signal drives the NPN transistor's base more positive, its collector-to-emitter resistance decreases. This places more negative voltage on the PNP transistor's base. The collector-to-emitter resistance of that PNP then varies according to the incoming signal, and changes the amount of current flowing through the audio transformer's primary winding. Through inductive coupling, the signal transfers to the transformer's secondary and causes the speaker to emit sound waves.

Now let's assume the amplifier's volume control (R4) is decreased and follow the action. The 5-k ohm potentiometer's wiper is moved toward ground, applying the signal across less resistance while increasing the wiper-to-base resistance "seen" by the signal. This reduces the amount of signal applied to the PNP transistor's base, lowering the volume. At the same time, the negative (forward) bias flowing through the NPN transistor to the base of the PNP transistor is reduced, biasing that device toward cutoff. The PNP transistor responds by increasing collector-to-emitter resistance and reducing the flow of battery current reaching the speaker. This dual action extends battery life when using low volumes. To see a real-life example, interrupt one battery lead on your handheld transceiver and monitor the current drawn while using different volume levels. The results are usually most enlightening.

Diodes and integrated circuits (ICs), are two additional components that are "spinoffs" of transistors. Simply explained, a diode is just an emitter-base junction of a transistor that conducts current in only one direction. This action can be used to rectify AC to DC or to detect RF signals, for example. The diode may also be used as a switch that's closed in one direction and open in the other direction. This means that the diode will present very low resistance from anode to cathode, but a very high resistance from cathode to anode. The exact ratio of these resistances is known as a diode's front-to-back ratio. Diodes are frequently used as an electronic switch by varying the voltage applied between their terminals within precise limits so the device exhibits one of two stages: cutoff (open) or saturated (a dead short). The sharp "knee" at a diode's point of saturation is also the underlying principle of zener diodes used to regulate voltages and to provide accurate voltage references. The zener diode is operated with its anode connected to negative and its cathode connected to positive voltage. This reverse biasing places the diode near its point of conduction. When the Peak Inverse Voltage (PVI) is exceeded, the diode begins to conduct; however, a special production process, which controls the silicon's resistance and places the zener point at the desired voltage level, prevents the device from being destroyed by overload.

One of the most significant innovations in electronics is the IC, which has become very common in most amateur radio equipment. ICs fall into two general categories: digital and analog. Digital ICs are used in frequency counters, memories, computers, and other areas adaptable to digital logic level of "1" and "0" (the binary system). Analog or linear ICs are used in signal applications like audio amplifiers, FM demodulators, etc. Transistors are the primary components of the internal design of both types of ICs. Some of the transistors are used as only resistors, some as diodes, and some as complete transistors. Additionally, small capacitors between base-emitter junctions are often employed as variable capacitors. A large number of transistors may be included in a single IC and thousands may be included in a single Large Scale Integrated (LSI) Circuit. Although one IC is a single component, its functions can be quite complex.

This chapter has been a fundamental discussion of basic electronic theory presented in plain language. It's not intended as a complete course in electronic technology or as a study guide. It's simply my way of trying to make complex electronics concepts understandable. I hope you found the information beneficial and that it helps you understand not only your radio but also the projects described in this book. And now it's on to the wonderful world of homebrewing!

CHAPTER 3
Projects for the Home Station

This chapter contains a variety of simple yet interesting weekend projects which can expand and improve your home station setup. Each one was selected to increase your enjoyment of our great hobby and to teach you many time-proven, helpful "tricks of the trade." You can mix and match projects as desired, but I encourage you to read this chapter in its entirety because it's loaded with good advice you'll use for many years to come. I also invite you to think of our discussions of upcoming projects as a visit with the author via the printed page. I say that because I consider this book a unique means of describing many shack "goodies" that often go unmentioned in brief on-the-air contacts. A local amateur may tell you he or she is using a KenYaeCom 5000 transceiver, a matching power supply, or perhaps an ABC antenna tuner plus homebrewed dipole, for example. But an actual visit to that same shack might reveal "sideline" items like a master control box, a unique microphone or keyer, or even a special project that would be ideal for your own station. We all grow through such "show and tell" endeavors, and that's precisely the objective of this chapter.

As an extra treat, the latter part of this chapter describes some easy ways to do a quick rework of older type ham gear often available at very low cost from hamfest fleamarkets. Now, before you start jumping with joy and congratulating me for showing you a "shoestring approach" to getting on the HF bands, I must emphasize that such "get by" measures can't compare to using a nice new transceiver. Budding newcomers need all the positive encouragement possible right at the beginning, and starting with less than you can honestly afford may unnecessarily limit your enjoyment. If you're a youngster on a zero budget, however, I (more than anyone!) understand your choice may be old-time gear or nothing! My first three or four "stations" consisted of discards acquired almost for the asking, but I could barely communicate out of the backyard. My best DX for several years was Puerto Rico, and that station lost me before the QSO ended. Hams can't survive on enthusiasm alone; they must also have good radios and good supporting equipment!

That's enough "K4TWJ philosophy": now let's get rolling on the various projects in this chapter. As you'll notice, many of these items are adaptations and modifications of presently existing items; others are "build from scratch" projects. This combination provides a good starting point, as well as room to develop skills, for just about everyone and every station. Good luck, and enjoy the projects!

Power Strip with "Big Switch" Simplifies Station Wiring

Many amateurs start with a single transceiver and power supply plugged into a convenient AC outlet and then add various accessories. Within a short time, the station has three or four AC line cords and a multiple outlet adapter/AC cube to handle the power hookups. When another "must have" goody is added, lack of sufficient outlets starts one thinking about extension cords. Extensions are easily added, but the resultant "plugging in" and "unplugging" tasks quickly become a nuisance. Simply leaving units connected and switching them on/off via front panel buttons isn't really a viable solution. Some items, such as battery chargers and TNC power adapters use wall-mounted AC supplies and must be unplugged to fully "switch off." Additionally, leaving an expensive transceiver plugged into an AC outlet while you're away from home can invite unnecessary (and avoidable) complications. Surprise electrical storms can always develop and, although the rig's power switch is off, can damage sensitive internal circuitry that's susceptible to lightning "spikes." Is there an easy way to sidestep all of these little pitfalls in a "one shot fix?" Yes indeed!

During the old days when ham setups could fill a whole room, many big-time operators installed a multiple AC outlet box with a large (actually huge!) master switch on the wall so everything could be switched

Figure 3-1. A multiple outlet AC strip adds always-appreciated convenience to any setup. See text for a discussion on selection of types and current ratings plus choice of home outlets.

on or off with a heavy "clunk." Such measures are a bit drastic today. There's an easier, but just as effective, method you can employ with a device that's as close as your local electronics store or amateur dealer. A multiple outlet AC strip with a built-in master switch like that shown in Figure 3-1 is ideal for most amateur setups in the "barefoot rig" category. In other words, these medium weight strips are fine for distributing voltage to a 100-watt output transceiver with accessories, and even a 45-watt VHF transceiver, simultaneously, but they are not suitable for a high-power linear amplifier's AC line. That's not a problem for our Novice/Tech friends, however, as their authorized power limit is 250 watts (isn't that convenient!).

So, how do we properly install the power distribution strip? First, let's determine which home outlet will be best to use. You need a modern three-prong outlet located close to your setup, and preferably one connected to its own circuit breaker. At first glance, the dedicated circuit breaker may not seem practical, but you can always "fudge" a little here. For instance, AC outlets in bedrooms are often connected to their own circuit breaker and used to power only a couple of small lamps, AM/FM clock radio, and the occasional TV. Such a circuit breaker's combined load is light, and may be close enough to having a separate line for your rig. After switching circuit breakers on/off and determining the best outlet, connect your multiple outlet strip and plug amateur gear into its sockets. Typically, you might have a VHF transceiver's power supply, HF transceiver's power supply, rotor, microphone and/or keyer supply, outdoor preamp, or other accessories plugged into the strip. You can then mount the strip on the side or back of the desk and use its built-in switch to "fire up" the full setup at once. Nice!

Figure 3-2. This "magic length" extension coax cable is easy to assemble and is a great asset when temporarily moving your rig into a den, setting up a portable station, etc.

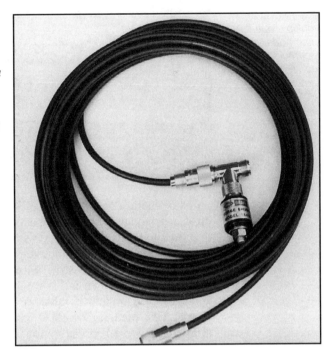

Notice we didn't include the clock in connections to the power strip. Naturally you want that item to stay on all the time. But there's another point worthy of consideration, too. Remember those lightning spikes I mentioned earlier? Avoid mounting your clock on top of your amateur equipment whenever possible. I've heard of many instances when lightning-induced spikes affected equipment even when it was unplugged from the wall, simply because the spike radiated from a clock on top of a rig. Such occurrences aren't common, but an inch of prevention is always worth a pound of cure! Also, you do unplug the antenna's coax cable from your transceiver or antenna tuner when not in use, don't you? Good show: you're avoiding all unnecessary opportunities for problems. One other point: it's a good idea to use the on/off switch on your power strip when you're near home, but to unplug its cable from the wall when you leave home for long periods. You still have the desired one-plug convenience—you've simply ensured that storm spikes or unauthorized hands have absolutely no way to reach your rig.

Power distribution strips are available in various sizes and wattages, so let's talk about which unit best fits your needs. The most common strips are marked on the back with 120 V, 15 A. That means this strip is for use with 120-volt outlets and can handle up to 15 amperes. We can calculate that total power handling ability as follows: 120 x 15 = 1800 watts. "Wow—that's much more power than I'll ever use!" you might say. Not necessarily: a 100-watt output transceiver typically uses a 13-volt/20-amp power supply, which draws 4 to 5 amps from the 120-volt AC line. A VHF transceiver will probably draw 3 amps; a TNC will draw 1 amp; etc. If you tally the current requirements of all the equipment in your station, it will probably add up to 9 or 10 amps (assuming 120-volt AC). In this case, a 15-ampere multiple outlet is a perfect answer. If you have a smaller station, a smaller wattage outlet will suffice.

Finally, double check the circuit breaker connected to the wall outlet into which you plugged the multiple outlet strip. Is it also capable of handling 15 amperes? Yes? Good! You now have a safe and convenient way to activate your station with a single "big switch"—just like the pros.

Magic Length Coax Cable Extension Has Unlimited Applications

How many times have you wanted to be able to quickly move your HF transceiver into an adjacent bedroom or den for a casual operating stint or for chasing early morning DX while having breakfast in bed? Maybe you've wanted to carry the rig along with you on vacations, or just to be able to operate from the patio, but postponed out-of-shack operating because the antenna hookups were a bit too difficult or SWR seemed to change for the worse when you added extra coax cable. Such considerations are a natural part of amateur radio life, but here's a simple solution that works great in 90 percent of the cases. Good news like this definitely warrants sharing, so here are the details in a nutshell.

This project is a homebrewed 34-foot-length of 50-ohm/RG-8X coax cable you can use as a transmission line extension or a ready-to-use feedline for impromptu HF antennas. What's special about this cable? Through many years of working with rigs and antennas, I've found 34 feet the ideal length for numerous applications, and it actually seems to lower, rather than increase, SWR when used with or added to HF antennas of all types. When moving my rig into

Figure 3-3. Snap-on toroidal cores like the convenient MFJ "four pack" shown here are available from amateur radio dealers nationwide. These items are very useful for minimizing RF feedback.

the den and using the extension cable from the shack (which connects through 50 feet of coax to the outdoor antenna), the SWR actually becomes a bit lower. This also holds true for my beam, vertical, dipole, and Delta Loop. During weekend vacations, I use the same 34-foot coax cable with my Cushcraft R-5 vertical antenna system. I quick-mount the antenna on an outdoor balcony, connect the coax, and bingo—I have a ready-to-use multiband antenna system with low SWR.

Checking out a new antenna? The 34-foot coax cable again comes in handy for setting it up or checking the antenna's SWR before adding the final length of transmission line. If the SWR changes when you switch from the "magic length" to your new length, you can be confident that the antenna needs tuning.

Have you ever considered using your mobile HF antenna with an indoor rig during mini vacations? Once again, the magic 34-foot length comes in handy. Just add an 83-1J double socket adapter in your auto, route the cable into your make-shift shack for the weekend, connect the rig, and you're ready to get on the air with a good signal. Additional uses for this simple project are unlimited and will seem to appear almost daily after you have the cable ready. Build one. You'll like it!

This extension cable, rolled up and ready for use, is shown in Figure 3-2. The strange item on one end is a lightning arrestor and 83-1T "Tee" adapter. The coaxial lightning arrestor bypasses surges to ground and is a worthwhile addition to any shack. Alternately, it can be unscrewed to provide male or female connectors on the cable's end. I assembled the cable from RG-8X Marine Grade coax so it could better survive abuse, since Marine Grade cable is more rugged than conventional cable. It has an ultraviolet-proof outer jacket, extremely good shielding, tough inner insulator/dielectric, and it can be used in blazing sun, buried in ice, or even immersed in water without detrimental effects. Understand, though, that I'm talking about the cable, not its end connectors. These should always be well weatherproofed and waterproofed for permanent installations or protected by a plastic bag, etc. during weekend/impromptu outdoor use. Another point to keep in mind: the life of any type of coax cable is maximized by proper handling. As you roll it around in your hand and stretch it out, you'll notice a natural tendency for curling in one direction (opposite its kink!). You'll also notice that the cable lays flat when unrolled in the manner that a wheel rolls, as compared to "pulling it unrolled" as you would a string. Remember this tip when handling coax cable and you won't distort the spacing of its conductors, which will help maintain its condition of "newness"). I've found that, when rolling or unrolling cable from a fixed/standing

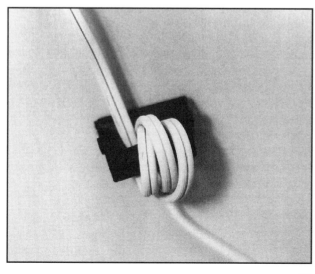

Figure 3-4. Quick-assembling a choke to minimize RF feedback and interference involves winding three or four turns of AC cable through a large snap-together toroidal core. Details in text.

Figure 3-5. Packages of 3-inch-wide copper foil are available from amateur radio dealers nationwide. They are ideal for ground strapping and other applications, as discussed in text.

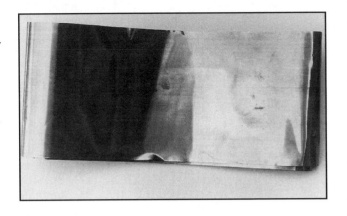

position, turning the loop around once or twice in your hand for each one or two winds of the cable corresponds well with its natural curvature and makes handling very easy.

The only additional step necessary for assembling this "magic length" extension cable involves properly installing PL-259 connectors. An in-depth description of this procedure plus information on installing other connectors is included in the Appendix of this book. I encourage you to review that information.

One brief reminder warrants mentioning at this point, however. When purchasing cable and connectors, double check to ensure the small coax "UG" adapter for the PL-259 is the right size for your selected coax. One size adapter fits regular (and slightly smaller) RG-58; another adapter fits (slightly larger) RG-8X. Check to make sure you have the proper adapter by sliding it onto the coax. Do not remove the cable's outer/rubber jacket. If the adapter slides on, it's the right size; if you would have to cut off the outer jacket first, the larger adapter should be used. Good luck, and may you find many good uses for this coax extension cable.

Eliminate RF Feedback with This Quick and Easy Fix

If you've ever experienced a sharp finger burn from the metal case of a transceiver or antenna tuner, noticed your packet system's TNC "hanging" on transmit, or received reports of distorted audio from a known-good transceiver, you've probably been the victim of RF feedback. This elusive phenomenon can affect any setup, especially those lacking a proper ground system. Amateurs who live in upper story apartments, etc. may not be able to install a proper ground, so here's a time-proven technique for minimizing RF feedback.

Before outlining the cure, let's discuss the cause(s) of RF feedback. Since most of us are familiar with audio feedback from high-power sound systems, let's use that as an analogy. If a speaker is pointed so that its output sound radiates directly into a microphone, the resultant audio feedback is a loud howl. Large objects in the area (like a blank wall) also reflect speaker sound into the microphone and produce feedback. Elimination in this case simply involves repositioning the microphone or speaker. RF feedback is basically the same phenomenon except that it occurs at your transmit frequency and is "heard" by your equipment rather than your ears. That latched-up vox, TNC, or continuously cycling automatic antenna tuner is thus being subjected to an excessive amount of RF energy from your transmitting antenna or other "phantom radiators." If your antenna is mounted closer than one wavelength from your rig, it may support RF feedback more than radiation of your signal. Ah, but you knew that when you selected an ideal site and installed the antenna, right? (Or is this a good time to reconsider your site selection?)

The most common cause of RF feedback is phantom antennas, such as resonant-length ground wires that pick up energy from your antenna and direct it back into the shack, the cable interconnecting your transceiver or TNC, a microphone cable, or even the AC line cord from a wall outlet to your equipment. Additionally, two (or more!) of the previously mentioned cables may work together as a phantom antenna. Determining "villain cables" can prove quite challenging to new and experienced amateurs alike. But there is a simple "fix" that works in 80 percent of the cases: you simply install snap-on toroidal cores onto various cables in your shack. These toroids, illustrated in Figure 3-3, are available in handy four-packs from amateur dealers nationwide. You simply pop open the toroid, wind three or four turns of cable through it, and snap the toroid shut to produce a simple, yet effective, RF choke like that shown in Figure 3-4. The coiled cable and toroid core effectively "break up" RF resonance so stray RF energy can't affect equipment. Install a core on the AC cable to your transceiver, microphone cable, ground wire, and TNC-to-rig cable, and you've covered most bases in one shot! Should RF feedback persist, or if you experience "computer hash" affecting rig reception, additional toroids can be snapped onto the computer's AC cable, monitor-to-computer cable, keyboard-to-computer cable, etc. In fact, you can really "clean up your shack" (RF wise!) by installing snap-on toroids on all wires and cables! Always remember this easy solution for RF feedback problems, and bear in mind that it can be applied to mobile setups as well.

Installing a Ground System for Your Station

As most Elmers will readily agree, a good ground system is very important for your station's performance, safety, and smooth operation. It's one of the best weekend projects you can pursue, and materials required for installing a ground system are available from radio stores nationwide. A common station ground minimizes the potential of shock between equipment cases, helps reduce RF interference and feedback, often improves reception and transmission, and is necessary if you use a vertical or long wire antenna. But, newcomers may ask, isn't the ground connection included in modern three-prong AC outlets sufficient for a simple station setup? Not at all: it's basically a safety wire for home appliances and has minuscule benefit for your station's RF applications. You can, however, integrate that AC ground in your own ground system so all will work in tandem.

The main items required to install a ground system are a roll of wide copper strapping, small nails or tacks for attaching it to your desk, several feet of flexible ground braid or shield removed from old large-size coax, some alligator clips for making jumper leads, and a 4- or 6-foot ground rod for mounting outdoors. Some amateurs assume that regular (round) wire is suitable for station grounding, but that's not necessarily the case. Think back to your amateur license theory studies, and you'll recall that RF energy exhibits a "skin effect." It travels along the outer area of a conductor, and flat strapping exhibits five to 10 times more area than regular wire. Thick copper strapping like that used by broadcast stations for grounding is expensive and difficult to work with. There is, however, a good alternative for amateur radio use.

The 3-inch-wide copper foil shown in Figure 3-5 is

available in rolls of 15-, 25-, and 50-foot lengths from amateur radio dealers like Ham Radio Outlet. This foil is quite inexpensive and is approximately three times thicker than household aluminum foil. The copper strapping is, therefore, quite easy to work with (it can be torn by hand), yet it's rugged enough for even professional use. If you really want to get top performance from a ground-mounted HF vertical, use this type of copper foil for radials, and the antenna will "get out like a bandit"! If using it for radials, incidentally, you may wish to tear the foil lengthwise to produce double length at half width (1-$1/2$ inches). This economical step has a second advantage: it minimizes your work in laying a ground system, as each copper foil radial is equivalent to six regular wire radials.

An outline of the station ground system is shown in Figure 3-6. You simply tack a 3- or 4-foot section of copper foil out of sight along the back edge of your station's desk. Jumpers made by crimping and/or soldering alligator clips to approximately 1-foot lengths of flexible copper braid are then used to quick-connect each unit's metal cabinet or ground lug terminal to the copper strap. Flexible braid can be purchased from amateur dealers or removed from old, large-size coax cable. The only prerequisite is ensuring this flexible

Figure 3-6. Outline of a station ground system implemented using copper foil and flexible braid. Discussion in text.

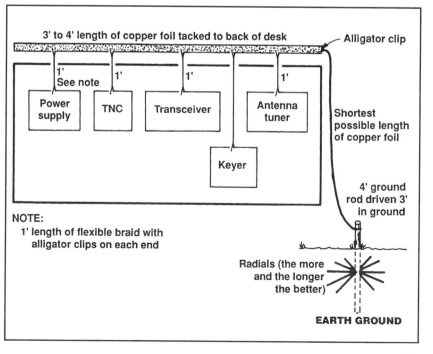

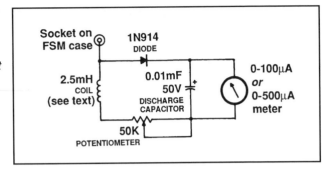

Figure 3-7. Circuit diagram of the easy-to-assemble field strength meter. Component values are not critical and substitution of working junkbox items is quite acceptable.

strap is as wide and as short as possible. Fold the strap at each end so the alligator clamp's end "folds over the strap" and holds it securely. Typically, you'll make and install flexible short jumpers for your transceiver, power supply, antenna tuner, TNC, and keyer. Do not rely on ground wires within interconnecting cables, but install a separate flexible jumper between each piece of equipment and the rear copper foil strap.

Finished? OK. Using an alligator clamp, now attach the shortest possible length of copper foil between your desk and the outdoor point where you'll install the ground rod. Use small tacks or nails to secure the foil in place, while striving to make that "run" as short as possible. Next, move outdoors and use a medium-sized mallet to drive the ground rod into the earth until only 1 foot remains above the surface. Watering the area with a garden hose before attempting to drive the rod in the ground will make your work much easier. Also, placing a 3- or 4-inch square of thick wood on top of the rod will make it easier to hit with a mallet and minimize damage to the rod. After the rod is in place, fold your copper foil as necessary and connect it to the ground rod. Remember to thoroughly weather-proof this outdoor connection. Your first-class ground system is now complete.

When adding new equipment to your station, remember to include a flexible jumper for connecting to the desk's rear ground strap. You may also want to add additional ground rods and buried radials of copper foil to this system to improve the performance of longwire antennas. You may also discover that the more you use copper foil, the more applications and benefits you'll find for it. For example, I've stretched it along wooden decks and used it as a ground plane for mobile antennas during weekend vacations. I've also made simple dipole antennas from copper foil laid on rooftops. This material even works well for wrapping bamboo poles to make elements for a homebrew beam. Think creatively and the applications are endless!

Quick-Check Your Transmitted Signal Strength with This Neat FSM

Do you occasionally wonder if your rig and antenna are pumping out their usual "like new" signal, or if your beam antenna still exhibits a high front-to-back ratio or gain over a reference dipole? Would you like a simple device for comparing performance of FM handheld antennas? Maybe you'd also like the convenience of a no-power-required meter for checking stray RF levels, or an independent device to confirm that your mobile setup is really working as the rig meters indicate. The simple Field Strength Meter (FSM) described in this project is the answer. In fact, daily applications for a good FSM are almost endless. You can use it for VHF, UHF, and HF, and you can slip it right in your pocket and take it anywhere. Attach a short wire to the FSM for an antenna, then watch its meter read upscale when you transmit. Rotate your beam antenna, and watch the meter reading change according to the antenna's direction. Connect another outdoor antenna to the FSM, and you can watch signal strength increase as the beam points toward it. Place the FSM across a room and then switch between your handheld's stubby ducky and $5/8$-wave antenna, and notice the difference on the meter. This little gem really eliminates guesswork and reveals the straight facts on what works best for you!

Ready to get started with your first big-time home soldering project? Great! The FSM's circuit diagram

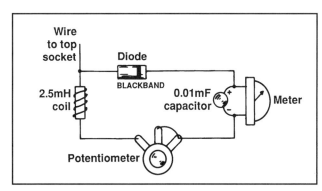

Figure 3-8. Pictorial layout of the FSM. Use this as a guide for interconnecting components.

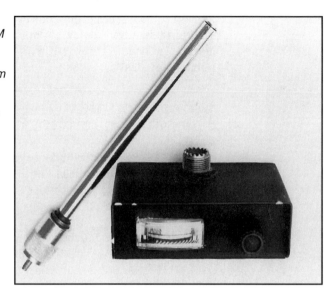

Figure 3-9. My homebrewed FSM is enclosed in a simple black cabinet made from sheet aluminum. The meter was salvaged from an old SWR bridge. The pick-up antenna was assembled by using a broken and restored pull-up antenna inserted in a PL-259 connector.

is shown in Figure 3-7. And, as an extra aid to our new amateur radio friends, a pictorial diagram which shows how to interconnect each component is provided in Figure 3-8. Simply put, the FSM is a miniature, "bare bones" receiver connected to a meter which indicates the strongest signal in the vicinity of its pick-up wire or mini-antenna. A signal is received by the pickup wire, roughly tuned in by the coil, converted to a miniature DC voltage by the diode, and applied to the meter. The potentiometer is used to set sensitivity and avoid "pegging the meter" when a signal is strong. The FSM can be assembled "open air style" in a small box or cabinet, or you can mount the four main components on a perf board and connect antenna and meter leads from the board to those components. While assembly and packaging is extremely flexible in this project, for best results, I strongly encourage you to use a metal enclosure.

To help you select an enclosure and layout for the FSM, my own homebrewed unit is shown in Figures 3-9 and 3-10. Sharp-eyed readers will notice that I made liberal use of components available in just about any junkbox. The meter, for example, was salvaged from a defunct SWR bridge I purchased at a hamfest fleamarket for 25 cents. I'm not sure if it's a 0 to 100 microamp, a 0 to 200 microamp, or a 0 to 500 microamp movement, but that really isn't important—any meter in those ranges will work fine. Remember, too, you can vary sensitivity of the FSM by its potentiometer. It may seem strange to read field strength on a meter showing SWR, but that's OK: it's simply a reference for comparison. The higher the meter reads, the greater the signal strength. I should also mention that you'd probably be more fortunate to acquire a round panel meter, whereas my hamfest "find" was a horizontal edgewise-type meter. I mounted a standard SO-239 socket on top because it accepts inexpensive banana plugs for short pickup wires, or I can plug one of my other antennas into the FSM for in-shack use. Additionally, a BNC-to-PL-259 adapter permits using FM handheld stubby ducky antennas directly with the unit. For use with HF gear, I also added a 19-inch pullup antenna, soldered to a PL-259 socket that can be plugged into the FSM's top socket.

Looking inside, the 2.5-mH coil and 1N914 diode are mounted on a small piece of perf board wedged in a corner. The .01-mfd/50-volt capacitor is mounted directly across the meter's terminals, and the 50-k potentiometer is mounted beside the meter. The 2.5-mH coil was an on-hand item, and many other types of chokes can be substituted. Alternately, you can wind your own coil using No. 18-26 wire and a popsicle stick or straw as a form. Wind 50 to 100 turns, then secure each end with a drop of glue. Remember to scrape the insulation from each wire "pigtail" before

Figure 3-10. Inside view of FSM showing location and layout of parts.

Figure 3-11. Electronic keyer with transceiver remote control was made by modifying an MFJ-422X "Combo Keyer." The switch on the rear tunes rig up or down in frequency, and the rear-mounted speaker serves as an extension for rig.

soldering. The .01-mfd capacitor and 50-k ohm potentiometer are also flexible, and any value between .01 and .001 mfd or between 25 k ohms and 100 k ohms will work fine. Other diodes can be substituted for the 1N914, but I prefer this component because it works well at both HF and VHF frequencies.

Take your time when assembling this FSM, double check wiring after soldering each connection, and it should work right off the bat. Notice that two wires connect to the 1N914: one goes to the antenna socket and one goes to the coil. The other end of the coil connects to one of the three terminals (either side) of the potentiometer. The center and other side terminal of the potentiometer connect to one meter lead (and its .01-mfd capacitor). Be careful not to damage the fragile little glass diode by overheating with the soldering iron. How? Flip a rubber band over the handles of your needlenose pliers so they clamp shut, then clip them onto the wire from the glass diode so the heat is dissipated by the needlenose pliers rather than reaching the diode Finally, paint the FSM's enclosure to match your rig, add a complimenting knob, then admire your handiwork.

One of your first applications for the FSM will probably be checking your home station's transmitted power level. Place the FSM on a convenient shelf in your shack, plug in a 3- or 4-foot wire, then adjust the front sensitivity control for mid-scale meter reading.

Record the potentiometer setting and sensing wire used for future reference. Perform that same measurement a few months later to ensure that everything is still up to snuff. You may find your rig is still delivering rated output and the antenna SWR is still low, but power has decreased. What happened? Quite possibly water seeped into the station's coax and increased losses. Without an FSM, you might not even realize that! The key here is to record potentiometer settings, meter readings, and pickup antennas used, and to retrieve those records later. Much of my VHF/UHF activities involve OSCAR (which stands for Orbiting Satellites Carrying Amateur Radio) satellites, and, since my 450-MHz transverter and power amplifier are solid state units without front panel meters, this FSM is an absolute blessing. I check 450-MHz signal strength by plugging the coax cable from my adjacent 144-MHz antenna into the FSM. The meter typically reads "2" when using the barefoot/10-watt transverter, and "8" when switching on the 100-watt amplifier. SSB microphone gain is then adjusted until the meter reads approximately one-half of full output, or "4." The FSM is also handy when assembling home QRP transmitters: I just watch the meter to make sure the homebrew circuit is producing output. You'll surely find many more uses for this flexible item. Good luck, and may your rig's field strength always remain strong!

Deluxe CW Keyer Includes Transceiver Remote Control

Heads up, late-night CW aficionados—our next project adds some real excitement to your Morse activity and it is a gem! I'm considering producing it commercially sometime in the future (yes, this is a "K4TWJ original"!), but I couldn't resist sharing its details with you now. The idea was inspired by my sincere enjoyment of CW and my desire to operate my rig from an adjacent room or the rear seat of an auto via a single remote cable (yes, CW mobile!). I began by adding an extension cord to my keyer and increasing the volume of my transceiver, then additional elements fell into place. The result was a combination keyer and paddle with built-in extension speaker for the transceiver plus up/down frequency tuning via an extra switch. Everything is thus combined in one palm-size control unit you can prop on an armchair or bed corner. The

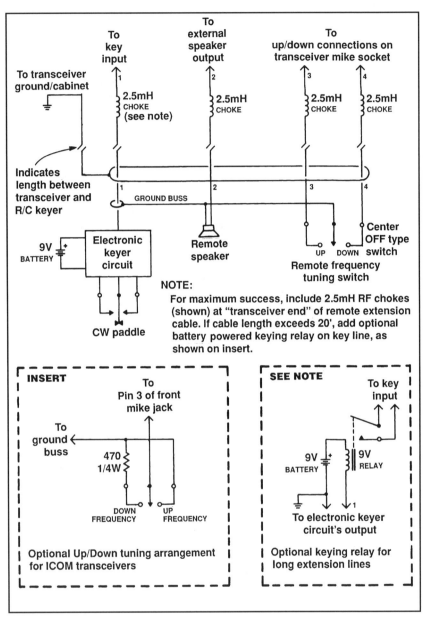

Figure 3-12. Wiring schematic for the remote control keyer. A small 4-conductor cable approximately 20 feet long carries the key line speaker and up/down tuning connections. All wires share a common ground/shield.

project is quite inexpensive, easy to build, and something you'll enjoy for many years.

This remote-controlled keyer may be assembled in several different ways to mate with your rig and operating preferences. It can be built completely from scratch with an electronic keyer kit and paddle installed in your own enclosure, or you might combine it with a ready-to-build keyer on a pc board. Alternately, you might expand/modify an existing commercially made keyer to save the hassle of drilling and mounting components. I chose the latter route and modified an MFJ-422X "Combo Keyer." Assembling the project took only a couple of hours, most of which time was spent visiting the local radio store for a four-conductor cable and microphone plug to match my transceiver. If you're "building from scratch," a good source of electronic keyer kits is Ten-Tec, 1185 Dolly Parton Parkway, Sevierville, TN 37862; Phone: (800) 833-7373. Small metal cabinets suitable for housing circuitry and a paddle are usually available at RadioShacks, but I leave mechanical details of construction to your own ingenuity.

A photograph of my previously mentioned adapted-from-commercial keyer with remote control is shown in Figure 3-11, and its rig-interconnecting/cabling scheme is shown in Figure 3-12. A small diameter four-conductor shielded cable, approximately 20 feet long, carries the keying line, external speaker wire, and up/down tuning connections. Since these wires share a common ground (the cable's shield), only a minimum number of wires is necessary. Check your transceiver's manual to determine if one or two wires are required for frequency tuning. ICOM transceivers, for example, use only one wire which, when connected to ground, sets the rig tuning up frequency. Connecting the same wire through a 470-ohm resistor to ground sets the transceiver tuning down frequency. Kenwood transceivers use separate wires connected to ground for tuning up/down frequency. Essentially, we're duplicating actions performed by up/down tuning buttons on a microphone. This information is included in the "microphone connections" section of your rig's instruction manual. You can quick-test this approach for remote tuning by double checking your rig's manual and connecting a short jumper between ground and the microphone socket's pin(s) required for up/down tuning. Try that now as a confidence builder, then continue reading.

Plugs mating with connectors/sockets on your transceiver are attached to one end of the remote control cable. These plugs should connect with the keyer socket, external speaker socket, and microphone socket (only frequency tuning connections are used on the latter). Connectors on the cable's opposite end should

41

Figure 3-13. Under-chassis view of a typical old-time receiver being restored to like-new. The large capacitors we're pointing to are filters to be changed to minimize hum. Just remove/replace one wire at a time, and the process will go smoothly. Discussion in text.

mate with sockets you add for up/down tuning, speaker connections, and keyer output. As an example, I removed the wires from an extra output on my MFJ-422X and used it for the up/down tuning connection while routing the speaker connection to a socket previously used for externally powering the keyer. Internal modifications to the keyer then consisted of adding a battery for stand-alone operation and rewiring the rear-mounted speaker so it acted as an extension speaker for my transceiver, rather than as a sidetone monitor for the keyer. Since all connections plug in, using the keyer at the rig or in an adjacent room (or even adding an additional extension cable) is a plug-in cinch. I must point out, however, that 2.5-mH chokes should be added in series with each lead (at the transceiver end) if extension cables longer than 20 feet are used. This measure sidesteps possible RFI problems.

Like any true amateur, I continued modifying and expanding this remote control keyer for bigger and better results. My first version, for example, used a single momentary toggle switch with center off for tuning. The toggle switch was attached with Crazy Glue to the keyer's rear panel. Since the photograph was shot, I've replaced the toggle switch with momentary pushbuttons mounted on a small "L" bracket glued to the keyer's rear panel. The pushbuttons permit much more accurate control of frequency tuning. My next step was remounting the complete unit in a larger case, adding a larger speaker for better audio quality, installing a single cable connector on the rear, and adding a rechargeable 9-volt battery for power.

Clever amateurs can take advantage of my development work, add their own expansions, and make a terrific finished product. You might also consider adding a voice synthesizer to your transceiver and routing its pushbutton-activation wire to an unused rear socket and on through the keyer cable. Assuming you add one additional pushbutton to the keyer, selected frequencies can then be voice-read back to you remotely.

Now visualize operating your rig with the remote control keyer. You can tune in other stations calling CQ, contact them using the keyer, vary speed as desired, monitor your own transmissions via the transceiver's internal sidetone (which feeds back to you through the keyer's speaker), and even operate full break-in, just like you were in front of the rig. Everything necessary for full-blown communications is right in your hand. If you'd like to open a new dimension in CW enjoyment, this project is a winner. What else can I say except get cracking on your own remote-controlled CW unit now!

Easy Restoration Breathes New Life into Old Shortwave Receivers

Many newcomers select a VHF FM transceiver as their first rig, then expand their interests into HF activities to the extant that their funds permit. The result is a quest for used HF equipment, and older model SSB/CW HF transceivers are now being snapped up as fast as they hit dealer shelves. Although many of these rigs are a real bargain, their cost can still prove an obstacle to those with limited pocket money for their hobby pursuits. But there's a really glamorous alternative that would make even well-equipped old-timers green with envy. The secret involves "mild-mannered "restoration of older-style shortwave receivers (and transmitters!) for CW operation on "low bands" like 80 and 40 meters. This approach can also be a lot of fun.

Many people restore classic cars from bygone days into beautiful vehicles that they use almost daily, and that same kind of romance can be recaptured in amateur radio! Yes, you can own and operate a genuine classic and enjoy amateur radio in style on a limited budget! Talk about fun!

Amateur transmitters and receivers of 1950 and 1960 vintage are often tucked away in attics and basements and "discovered" through conversations with old-time hams on 2-meter FM. I've heard of several occasions when budding newcomers acquired like-new receivers and/or transmitters complete with manuals and packed in their original cartons. A close friend acquired a marvelous National brand hamband-only receiver through a 2-meter conversation: the only stipulation was that he remove all the equipment from the basement in return. In the process, he found several antenna switches, an antenna tuner, power supply, and a key that was still in its box. Another good source of shortwave receivers and transmitters is hamfest fleamarkets. Try to enlist an Elmer to help you select and check out equipment before you purchase. Or simply ask the seller to plug a unit that you're interested in into an AC outlet to confirm proper operation. I've found several terrific rigs in hamfest fleamarkets. My most impressive finds were an almost-mint Johnson transmitter that I purchased for only $5.00 (and this 50-watt rig works great!) and a nice Hallicrafter receiver I obtained for only $15.00.

When shopping hamfests, incidentally, remember that the early bird gets the worm. Try to arrive before the doors open and notice the equipment as the folks unload prior to setup. You may pay a few dollars more for dealing immediately, but you can get a magnificent piece of equipment in the process.

Once you understand the shortcomings of classic equipment, the situation can be reversed in your favor. Forty-year-old receivers typically "drop off" in sensitivity on bands above 20 meters, but still work fine on lower bands. Novice/Tech licensees are authorized to operate voice/SSB on 10 meters, but activity on this "daytime band" is presently poor as a result of the low sunspot count. Meanwhile, Novice/Tech CW activities on "evening bands" like 80 and 40 meters is increasing. Uncomfortable operating CW? You can upgrade your packet TNC to a multimode controller, and let it help you read incoming Morse code, while you simply type outgoing transmissions on the (interconnected) computer's keyboard! Plus, operating on the "low bands" is the perfect way to contact those nearby states for your Worked All States (WAS) award chase. Need more inspiration? Simply visualize the soft amber light of genuine radio dials and the sparkling array of tubes glowing in a dim room. Friends, this is amateur radio at its best!

If you have access to a pair of diagonal cutters, can solder a couple of wires together, and plug in a microphone connector with its little "locating key," the restoration of old-style shortwave receivers will be a cinch. Begin by connecting a short wire or outdoor antenna to the unit, switch it on, and tune in a few signals to check performance. You'll probably notice some hum coming from the receiver's speaker. Disconnect the antenna and see if the hum's level remains constant or varies as you increase receiver volume from minimum to maximum. If the hum level remains steady, it's probably due to defective (dried out with age) filter capacitors. If the hum level changes with volume settings, replacing bypass capacitors usually restores normal performance. If receiver sensitivity seems low or speaker volume minimal, a simple tube replacement will probably solve the problem. Likewise, a burned-out dial lamp or mouse-chewed speaker cone can be replaced with a simple component-for-component swap. After completing these repairs, the cabinet, knobs, and dial can be cleaned with a damp cloth and window cleaner to produce a sharp-looking finished product.

Figure 3-14. Replacing a tube in an old-time shortwave receiver. Pay close attention to the direction of the locating key (in center of base's circle of pins), then reinsert new tube accordingly. Replacing tubes in old-time receivers makes a significant improvement in performance.

After your initial evaluation is complete, unplug the receiver from its wall outlet and wait a few minutes for any possible residual voltage/charge in capacitors to dissipate. Next, remove its bottom plate or cabinet to access the internal parts for replacement. Study the under-chassis components and locate the two or three largest tube capacitors (Figure 3-13). These items are located in the power supply section (near the power transformer and/or rectifier tube) and are usually marked with 10 (or 20 or 40) mfd, 150 (or 250 or 450) VDC. One end of the capacitor will be marked with a black band, indicating its negative terminal. In some cases, two capacitors may be enclosed in one case. These components have two positive leads emerging from one end and a single negative lead emerging from the opposite (silver) end. Jot down values of all filter capacitors as a "shopping list," then notice smaller wax-coated or plastic tubular-type capacitors in the receiver. Their values will typically be .1 mfd to .5 mfd @ 150 to 350 VDC. Notice these bypass capacitors are also marked with a (minus) ring on their negative end. Add these components values to your shopping list.

Now turn the receiver over and look at its above-chassis components. Tube replacements will depend on initial receiver performance, the desired level of restoration/performance, and available funds. Tubes are typically marked with numbers like 12AX7, 50C5, 6BA6, etc. Remove a tube by gently pulling at its base and then notice the positions of pins and/or center locating "key" before setting the tube down. A tube installs only one way to minimize chances of mistakes. Installing a few new tubes is easy, and it often breathes new life into old receivers (see Figure 3-14).

A sharp thinking and fast acting newcomer can purchase a receiver during opening hours of a hamfest, find a quiet spot to pop open its case, make an on-the-spot shopping list of capacitors and tubes, then return to the fleamarket area and purchase those components at very low cost. You may also be able to find the parts you need at a local radio supply store or through mail order.

Here are just a few additional tips for component shopping. If a receiver is over 15 years old, assume that its filter capacitors need to be replaced, regardless of obvious hum. You don't need exact value components for replacement: simply remember to use larger than original values. For example, a 10-, 20-, or even

40-mfd capacitor is fine for replacing an 8-mfd capacitor (assuming equal or higher DC voltage rating). Also, "2-in-1" capacitors like 8-mfd and 20-mfd can be directly replaced with separate 20-mfd capacitors (often less expensive in pairs at hamfests). Likewise, in bypass capacitors, .1-mfd capacitors can usually be substituted for values like .05mfd. The negative leads from both filter and bypass capacitors usually connect to the receiver's ground/chassis (a convenient means of confirming that you've spotted the right component).

Now that you have the right parts, be sure the receiver is unplugged from an AC outlet, then replace the capacitors on a direct wire-for-wire basis. Cut the red/positive wire first, bend the capacitor back, then solder the wire from the new capacitor to the terminal/connection where the old capacitor was attached. Next, clip the old capacitor's negative lead and solder the wire from your new capacitor to that point. Work as neatly as possible, don't leave long lead lengths that can fall against other wires and short circuit, and the restoration process should go smoothly. Just remember to replace one capacitor at a time, lead-by-lead.

Final restoration involves changing tubes, cleaning the inside (and outside) with a soft bristled brush and mini vacuum, and maybe adding a nice coat of automobile wax if the rig has a smooth-finish metal cabinet. Some famous names of receivers that stand proud today include Hammarlund, Hallicrafters, National, and Collins. Finally, I encourage you to hold onto your restored pride-and-joy, even after you acquire a fancy new transceiver. Like classic autos, these delightful old radios are becoming scarcer by the day and are continuously increasing in value. Enjoy owning a genuine classic!

Rejuvenating Golden-Oldie Transmitters

The next unit needed to complete your super-economy station is an old-model transmitter, and these rigs are also frequently discovered in basements or hamfest fleamarkets. Some are quite large and heavy, some are quite small, and most produce at least 40 watts of output power. Some popular manufacturers of transmitters were E. F. Johnson/Viking, Heathkit, Harvey Wells, and Hallicrafters. Browse through amateur radio magazines of the 1950s and '60s (often filed in the

Figure 3-15. E. F. Johnson "Adventurer" transmitter found at a hamfest and purchased for only $5. Restoration took only a few minutes' time, and the little gem works like new!

archives section of libraries in larger cities) to become familiar with these transmitters. You can then spot the real thing at hamfests. One of my own special finds, the Johnson transmitter previously mentioned (and purchased for only $5!) is shown in Figure 3-15.

The same guidelines discussed for finding and restoring receivers applies to transmitters. If possible, try to check out a unit before purchasing it, and get assistance from an Elmer. Vacuum tube transmitters use higher voltages than receivers, so remember the following safety tips. Be sure the transmitter is unplugged from its AC wall outlet and left switched "off" for at least 15 minutes before removing its cabinet. Bleeder resistors should discharge capacitors during that 15-minute period. If internal tubes have a top-mounted plate connection, use a flat blade screwdriver to temporarily short that connection to the chassis (only when the rig is off!). If you see a small spark, it indicates the bleeder resistors (connected in parallel with filter capacitors) are defective and should be replaced with 500k ohm resistors of the same size/wattage as the originals. If a spark isn't apparent, the bleeders are working properly and shock hazards are minimal.

After removing the transmitter's cabinet, look for any charred components that may need replacing. Check the tubes: their glass envelope and base often become loose due to high temperatures when transmitting, which can worsen with aging. Basic-style transmitters typically use only three or four tubes, and they may still be capable of producing nearly full output. Still, your best bet is complete tube replacement —you can then keep the original tubes as spares. Additionally, your transmitted signal will sound like it's produced from a beautiful new rig. Furthermore, there's a unique "sparkle" to the sound of tube transmitters compared to transistorized transmitters/transceivers. In other words, tube replacement gives you a terrific on-the-air sound. Figure 3-16 illustrates tube replacement in progress.

Carefully turn the transmitter upside down and use wood blocks or books to stabilize its position. Be sure the transmitter is unplugged, then replace the power supply's filter capacitors as discussed in the section on receiver restoration. Tubular filter capacitors used in transmitters are usually larger than those used in receivers, in mfd rating, voltage rating, as well as physical size. Typical values range from 20 to 60 mfd @ 500 to 700 VDC. These items may still be good (as indicated by little or no hum on your transmitted signal), but their electrolyte has dried with age and they're prone to "popping" after a few hours use.

Capacitors of equal or larger mfd and voltage rating can be substituted on a wire-for-wire basis, also as described in the receiver restoration section. Watch those polarities! Remember that the black ring on the capacitor's silver end designates negative (and connects to ground). Remember, also, to ensure short wire lead lengths when installing new capacitors. Long leads are prone to falling against other circuit wires and short-circuiting. Again, I urge you to work slowly and accurately. Make sure that each capacitor junction that's affected by the replacement is soldered firmly and that the solder doesn't "bridge" between adjacent terminal strips.

One additional tip: if you can't find filter capacitors with the required voltage ratings, consider "making them" from available capacitors. For example, since voltage and capacitance divide in parallel, two 40-mfd, 450-VDC capacitors hand-wired in series produce a single capacitor of 20 mfd, 900 VDC. The two capacitors can then be used to replace a single capacitor in the transmitter. The only disadvantage of using two instead of one is that they require double the space, but extra room is usually plentiful in old-time rigs.

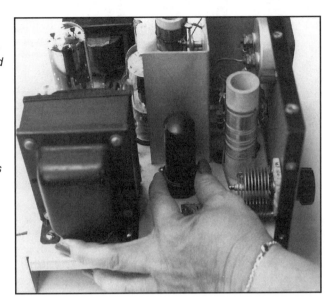

Figure 3-16. Replacing a transmitter tube is best accomplished by holding and slightly rocking its base while applying upward pressure. This simple change ensures that your classic rig delivers a great signal.

Your classic transmitter should now be performing admirably after its "electronic tune-up." Now use a soft bristled brush and mini-vacuum to clean the inside. If the transmitter is crystal-controlled rather than VFO-controlled, you might also consider adding an old-style (open air), three section 365-pfd AM radio broadcast tuning condenser in series with one of the crystal sockets' wires. This frequency-warping trick will allow shifting the crystal's frequency a few KHz to either side of its center frequency. Connect a stiff wire between all sections of the tuning condenser to achieve the greatest capacitance. A three section capacitor, for example, will thus have 365 x 3 = 1,095 pfd total capacity. Check the transmitter's crystal socket. Notice that one wire connects to the inner circuitry and the other wire connects to ground/chassis. Since your added tuning condenser is mounted on the transmitter's chassis, reroute the wire from the crystal socket's ground lead to the stiff long wire you added to the tuning condenser. A schematic outline of this "rewiring" is illustrated in Figure 3-17 for your convenience.

Now replace any obviously burned under-chassis components, including pilot lamps, then very carefully check the transmitter for proper operation while it's out of its cabinet. If you're not familiar with high voltages and vacuum tubes, again seek out an Elmer's assistance. High voltage can cause quite a shock, RF energy causes burns, and touching hot transmitter tubes can cause blisters. I'm not trying to scare you: I simply want you to respect an open transmitter just as you would a radiator fan and drive belts in an auto when its motor is running and the hood is up.

A convenient way to check out a transmitter involves cabling its RF output to a wattmeter or SWR bridge connected to a dummy load or antenna. After the rig has warmed up for a couple of minutes, you should close the key and immediately tune the "plate tuning" control for minimum current. Power output should rise to maximum as plate current drops to minimum. The transmitter's "antenna loading" control can then be increased in small increments while the "plate tuning" control is readjusted for minimum plate current (and maximum output). Additional, and more accurate, tuning information is included in the transmitter's original instruction manual.

You can either pronounce your classic transmitter project complete, or you can consider going one step further by repainting its case in a close-to-original color with automotive "touch up" paint, available from discount stores nationwide. Now place your "new" transmitter beside its awaiting receiver, stand back, and admire a real ham station that lights up, warms to the touch, and increases in value with each passing year!

Simple Concept Gives Full Break-In Operation with Older Receiver/Transmitter Combos

Among the questions most often asked about using older transmitter/receiver combos are those about antenna switching. Each unit has an antenna connector, leaving newcomers to wonder if they must unplug the antenna's cable from the receiver and plug it into the transmitter (and vice versa) for transmit/receive switching. During the "good-ole days" amateurs purchased now-extinct coaxial relays or assembled their own antenna changeover relay in a metal box, with transmitter and receiver on one end and antenna output on the other. But a much simpler solution is right at hand. It's especially attractive for "low band operations" and produces full break-in operation just like modern deluxe-style transceivers.

Several interrelated factors underlie this project. First of all, signals on lower bands are usually quite strong.

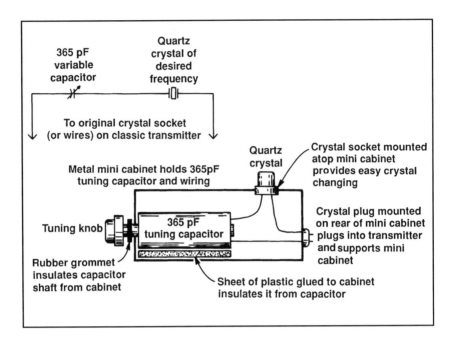

Figure 3-17. Simple crystal-warping circuit used to shift frequency several kiloHertz. This concept is much less expensive than a VFO and gives you reasonable frequency agility.

Older receiver performance on these bands is usually outstanding, and most old-style transmitters only produce output when their CW key is closed. Building on this knowledge, you can achieve good performance by connecting your main transmit antenna directly to the old-time transmitter and connecting a separate single 30- or 40-foot-long wire to the receiver. You then insert small diodes between the receiver's antenna and ground connections to avoid "front end" overloading when transmitting. The receiver's Automatic Gain Control (AGC), if so equipped, is then set to "fast," and T/R switching is eliminated. The receiver operates continuously, so you can listen "in between" transmitted dots and dashes if you want. Additionally, the receiver can be used for transmit monitoring (side tone) by simply turning down its gain while sending code.

Does this approach really work? You bet! I've been using it with my classic rigs for many years, and it adds a neat flair to Morse activities. Plus, I can often copy DX stations better than others because my receiving antenna is laid on the ground (honest!) and picks up very little noise. Ah yes, another secret of low band communications put to use. Low noise levels, rather than high receiver sensitivity, makes the big difference on 80 and 40 meters. Now let's discuss integrating this project with your transmitter/receiver.

An outline of the dual antenna setup is shown in

Figure 3-18. The transmit antenna can be a dipole, Delta Loop, or any antenna of your choice and is connected to your transmitter using a standard PL-259 plug. The receiving antenna is a random wire between 30 and 90 feet in length according to your preference, available space, supports, etc. Regular insulated "hook-up" wire between No. 14 and 24 gauge is suitable for this receiving antenna. Try to mount the single wire/receiving antenna at right angles to your transmitting antenna and as far away from it as possible. You can string the wire antenna along house eaves, under a garage, beside a wooden fence, or even lay it on the ground. Will an antenna laying on the ground pick up signals? Yes indeed! Try it! Just remember to increase your receiver's RF gain to near maximum as both noise and signal levels will be a couple of S units lower than normal.

Most old-style receivers use three rear-mounted screw terminals for antenna connections. A jumper may be included between one terminal and ground for connecting coaxial-type antennas. Connect your receiving antenna to the single/ungrounded screw terminal. If in doubt, check your receiver's manual or touch the antenna lead to each screw terminal and notice which one provides the best reception. Next, add a pair of back-to-back connected glass diodes,

Figure 3-18. Outline of dual antenna setup for producing full break-in operation with old-time transmitter and shortwave receiver.

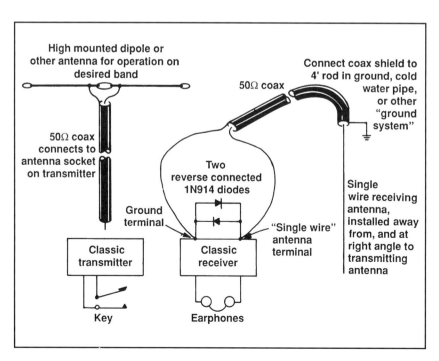

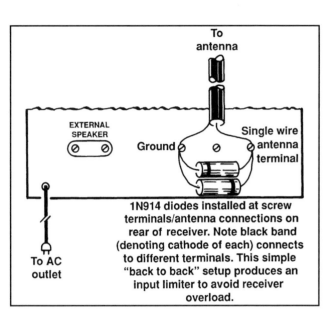

Figure 3-19. Arrangement for installing back-to-back glass diodes at antenna terminals of receiver. This simple arrangement prevents overload from the transmitted signal.

such as 1N914 or 1N34, between the screw terminal connected to the wire antenna and the receiver's ground terminal. These diodes will protect the receiver's "front end" from RF overload during transmit. An outline of back-to-back connected diodes is shown in Figure 3-19. Notice that the "banded" end of each one is reversed and that the diode leads are firmly twisted together rather than soldered. Diodes provide a .7-volt limiter and short higher power levels to ground. I suggest twisting leads rather than soldering to minimize the possibility of "burning out" the diodes from excessive soldering iron heat.

Your super-economy station with full break-in operation is now complete. The receiving antenna will work fine on both 40 and 80 meters, plus it will perform well for general shortwave listening and monitoring. Enjoy your "like-new" classic setup, and may the force of good HF DXing be with you!

This chapter outlined a number of semi-technical projects for the home station, and I hope you found them interesting. I tried to include a healthy amount of "Elmering guidance" in these project discussions to help you expand your knowledge and enjoyment of amateur radio. There were also plenty of useful tips and ways to apply knowledge, so I would suggest to you that you re-read this chapter in a year or so. Your insight into amateur radio will be even greater then, and even more homebrewing ideas will be apparent to you.

CHAPTER 4
HF Antennas

Surely the most popular area of homebrewing among radio amateurs is building antennas to use on the air. Why? Self-assembled antennas are very economical, easily configured to mount in a wide variety of locations, and are great examples of personal ingenuity. Knowledge of various antenna designs and the procedures involved in assembling and installing them also increases your confidence in communicating from almost any location. Want to enjoy some real amateur radio excitement? Get busy building your own antennas. It's terrific fun and extremely rewarding!

Approximately 50 percent of the new licensees I hear or talk with on 2 meters are anxious to get started in HF activity and diligently seek information on simple yet effective antennas for 10 meters. In too many cases, a would-be Elmer suggests unnecessary and discouraging shortcuts, such as using a backyard clothes line, random length wire, or a tuner to convert a rain gutter into a makeshift antenna. Or perhaps the budding amateur is encouraged to experience our exciting world of HF communications with only a dipole or ground-mounted vertical. While such a "bare bones" approach is several steps above our previously discussed compromises, an even more effective antenna can be home-assembled just as easily. I urge you to stretch as far as possible right from the start and build the best antenna you can fit into the available space. This is especially important if your rig runs less than 100 watts output. An exciting world of HF communications awaits you: don't shortchange yourself unnecessarily. When you walk outside, look up at your antenna, and feel a real sense of pride ("Wow, what an impressive antenna for such a basic indoor station!)," you'll know you're on the right track to getting maximum enjoyment out of our great hobby.

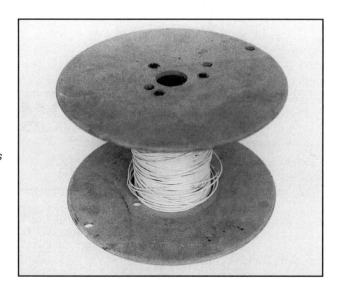

Figure 4-1. Hamfest fleamarkets are a good source of wire for homebrewed antennas. The heavily used roll shown here was acquired for only $1 and numerous antennas have been built from it.

The antennas described in this chapter represent a good cross section of designs that are easy to duplicate and offer better-than-average performance. Our main focus is on 10 meters, as it's the most popular HF band among Novice and Tech-Plus licensees. Any of the featured antennas, however, can be easily scaled for operation on other bands like 15, 20, 40, and 80 meters, or even WARC frequencies. Be sure to keep this book handy—I'm sure you'll want to refer to it often (and this chapter in particular) now and after your future license upgrades.

But First: A VERY Important Note

If you're a new amateur and are assembling your first antenna, make sure that you read the discussion of the 10-meter Delta Loop (the first project in this chapter), no matter what your antenna plans are. It contains a wealth of information and useful tips that will be applied to the other antennas we'll be discussing. There's a lot of common information that I state only once, in that section, rather than continuously repeating it, and that's where you'll find the details on working with coax, soldering connections, installing insulators, etc. We're going to build upon acquired knowledge as we go—this technique will help you better assimilate what you learn and also saves room so we can squeeze in more projects, and *that* assures that you get a book jam-packed with plenty of terrific information.

Having said that, now let's look at some general information about HF antennas, and then we'll move on to our particular projects.

Antenna Wires and Insulators

The material you use to make an antenna is, to a large extent, a matter of personal preference. Aluminum tubing is popular for verticals and beams because it's lightweight. However, 10-foot sections of less expensive (and heavier) electrical conduit is suitable for 10-meter antennas. Probably the most common material for antennas is stranded, enamel-coated wire available from dealers nationwide. Be sure to scrape the enamel off before making connections to this wire (a pocket knife works fine). Stranded copper wire, incidentally, tends to turn black with age, although it will still work fine. Another popular wire is copperweld, which consists of a steel core with copper coating. Be careful when using this strong, solid wire: it's very tough and can recoil in a flash, possibly inflicting sharp cuts.

Regular "hook-up" wire of No. 14 to 18 gauge is another good choice for antennas and is usually available at a very attractive price. Additionally, plastic insulated hook-up wire is easily dipped in blue/gray paint for a camouflage effect that makes an antenna almost invisible against the sky. The now-almost empty roll of hook-up wire shown in Figure 4-1 was acquired at a hamfest fleamarket several years ago (for just $1!), and has been used to build many great wire antennas. May you be equally fortunate finding such a bargain in the future!

Finally, ultra-thin "hook-up" wire between No. 20

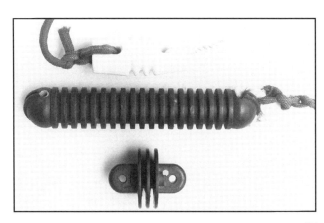

Figure 4-2. Antenna insulators are readily available in a wide variety of styles and shapes from amateur radio equipment dealers nationwide. Selection is mainly a matter of personal choice— they all work fine!

Figure 4-3. Some popular types of 50-ohm coax cable. Top (smallest) diameter cable is RG-174U which is rated at 300 watts and somewhat lossy. Next is RG-58/U and RG-8X. These cables are rated at approximately 700 watts, and RG-8X-type is low loss (preferred). Larger cable (bottom of photo) is RG-8/U (also made in 9913 and RG-213 versions, both of which are very low loss and high quality).

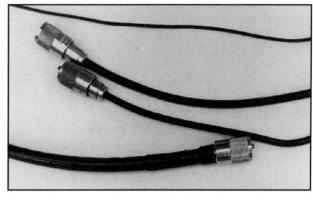

and 30 gauge can often be used for making "invisible" antennas. Indeed, this wire is so thin that it resembles spider webbing! The pitfall to using such thin wire is its tendency to break with stress. This can require that the wire be supported every 20 to 25 feet, again allowing slack for movement in the breeze. But ultra-thin wire often is the ideal solution for erecting an outdoor antenna in seemingly impossible situations.

Although some amateurs use wooden blocks and various other non-conductive material for insulators, the most common and readily available are the high-impact plastic or ceramic types like those shown in Figure 4-2. These insulators are available from amateur dealers nationwide at very low cost. Each style works fine, and the design is a matter of personal preference. The white "dog bone" insulators have a nice classic appearance and hold up well in any type of weather. More modern "thinned" and "extended length" types are not quite as obvious in the air. Just make sure you use insulators suitable for your particular climate.

Types of Coax Cable

The cable interconnecting your antenna and transceiver is your radio lifeline, and selecting the best type suitable for your particular application is very important to long-term results. Never shortchange this area by using old or "cut rate" coax. If in doubt, just select the most expensive type available! Also, avoid splices: a single length connected between your antenna and rig is always the best bet. Coax consists of a center conductor and an outside shield. When working with coax, be sure to avoid shorting those two conductors! Coax has a life span varying between seven and 10 years, so remember to replace it at the appropriate time. Figure 4-3 shows some of the popular coax cables available, and a basic discussion of the types most often used follows.

RG-58/U: Most popular type of 50-ohm cable. Suitable for HF applications when power is not over 700 watts and length of cable does not exceed 100 feet. Suitable for VHF applications when cable length does not exceed 25 feet. Slightly lossy.

RG-8/U: High power version of conventional RG-58/U cable. 50-ohm impedance. Suitable for both VHF and UHF applications. Handles up to 3000 watts of power.

RG-59/U: 75-ohm version of RG-58/U cable. Handles up to 1000 watts. Slightly lossy. Like RG-58/U, good for short length runs.

RG-8X: Improved grade of 50-ohm cable with low loss and very good characteristics. 50-ohm impedance. Handles up to 650 watts RF. Good for both HF and VHF applications. An outstanding type of "universal" coax. A marine grade-version of RG-8X has recently been introduced. This is good for all applications and has the same characteristics as RG-8X, except it's more immune to extreme weather condi-

Figure 4-4. "Coax Seal" is available in various size rolls from amateur radio equipment dealers nationwide. This product is exceptionally good for weatherproofing antenna connections and cables.

Figure 4-5. Antenna baluns are available in a wide variety of styles and are a very good investment. The one shown here was obtained from "The Radio Works," Box 6159, Portsmouth, VA 23703 and was used on a dipole for two years (hence the tape and Coax Seal smudges). The item is fully sealed, however, and still good for many more years of service.

tions, moisture, etc. If you're looking for a good small-size cable, this is it.

RG-213/U: Top quality cable with very low loss. 50-ohm impedance. Good for both HF and VHF applications. Handles up to 4000 watts of power. A marine-grade version of this very low loss and high-performance cable is also available.

When working with any type of coax cable, always ensure that its inner structure and both conductors are protected from weather. Moisture can seep into openings, such as those caused when you separate shield and center conductor into pigtail leads. Moisture can quickly be drawn along the cable's length by capillary action, rendering the coax almost useless. Be sure to seal all exposed cable ends and outdoor connections. One of the best sealers I have found is "Coax Seal," shown in Figure 4-4. Rolls of Coax Seal are available from amateur dealers nationwide.

Baluns

A balun transformer is used with many types of centerfed antennas and is applicable to many other designs. This device minimizes RF currents flowing in the coax feedline's shield, reduces TVI effects, and ensures a proper antenna radiation pattern. Instruc-

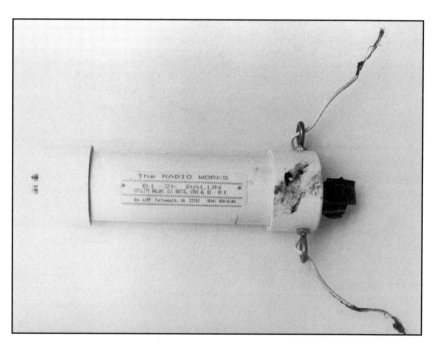

tions for installing and using baluns are usually included at purchase. A photo of a popular balun is shown in Figure 4-5.

Simply explained, a balun is a BALanced to UNbalanced transformer and is usually available in ratios of 1:1 and 4:1. The 1:1 balun is most often used in dipoles, and the 4:1 balun is used in antennas with 300-ohm impedance, such as folded dipoles. Numerous theories and discussions regarding the benefits of baluns fill amateur radio magazines each month and so will not be repeated here. I've personally found that a good balun helps an antenna's performance. Your choice of using a balun is strictly discretionary. One final note, however: some multiband antennas, such as the Carolina Windom, require the use of baluns, whereas vertical antennas seldom use a balun. (Additional information on antennas is included with discussions of various projects in this chapter, so read on and continue expanding your knowledge!)

An Easy "Armstrong" Method for Erecting an Antenna

If you're setting up your first HF antenna, try to mount it as high and as clear of obstructions as possible. For example, a low-slung horizontal antenna strung between a house eave and garage corner, or a vertical squeeze-fitted between a brick building and tall fence, lack "breathing room" and won't radiate effectively. Erecting a horizontal antenna between, rather than through, trees yields noticeably better results. Similarly, it's far preferable to mount a vertical on top of a roof as a ground plane, rather than right at ground level as a basic vertical.

One of the best techniques I've found for erecting antenna supports in between trees involves using a weighted tennis ball (see Figure 4-6). I attach a thin nylon or dacron rope to a large fishing weight, slit the tennis ball, and insert the weight inside. To the nylon cord, I attach a pull-up rope which will be used to raise the antenna. Then, using a glove for a good throwing grip and to avoid finger burns, I toss the tennis ball over a high tree limb in true baseball pitcher fashion. Within three or four tries, the tennis ball sails over my chosen tree limb, trailing its nylon cord and pull up rope. I retrieve the tennis ball, attach another pull-up rope, move to the next tree limb, and repeat the procedure. After all the pull-up ropes are in place, I attach them to the antenna, and simply pull the completed skywire up into position, tying off the ropes but leaving some slack so the antenna and tree limbs can flex in the breeze.

Using this method lets you keep both feet solidly on terra firma and only nylon rope has to get entangled in tree foliage. The antenna proudly swings in the open and there are no pull-up wires to interfere with performance. Lowering the antenna for tuning or maintenance simply involves untying the pull-up tie-off points. Remember to leave enough slack so the ends can reach the ground and you can still reach the nylon rope!

Generally speaking, your tennis ball-throwing ability will determine the height of your antenna. Consider 25 feet as a general minimum and 60 feet as a general maximum. Always pay close attention to power lines and neighbors' property when installing any antenna! Try to erect your antenna close enough to your transceiver so no more than 70 feet of coax cable is required for hook-up. Whenever possible, try to orient the antenna so it is not directly over your "shack" to minimize the possibility of RF feedback. Likewise, avoid mounting a vertical directly beside your station (so RF energy is directly radiated back to the rig). Be careful to avoid sharp bends when routing the coax cable to your station. Also make sure that its outer jacket is not pinched or cracked by tight squeezes through windows.

If you have extra coax cable after installing the antenna, cut off the excess rather than coiling it up on the shack floor. Excess cable length reduces the amount of power transferred to your antenna and may limit range. This doesn't mean your cable should be pulled tight, however. Leave enough excess cable outside the window to droop down toward the ground and make a wide loop before it enters the window (or you may experience a nice indoor puddle after a rain!).

Be sure to check your antenna's Standing Wave Ratio (SWR) before pronouncing it ready for operation. SWRs above 2:1 cause modern transceivers to reduce their RF output for safety reasons. The best way to lower a high SWR involves extending an antenna's length to lower its resonance frequency (or shortening its length to raise resonant frequency). Assuming you cut the antenna slightly long to start with, a simple "pruning" (by lowering each end's nylon rope) is all that's required. Before soldering the indoor PL-259 connector, you may also wish to prune the coax cable's length in 1-foot increments until you achieve a very low SWR (this step isn't vital; it simply ensures that your modern rig runs cool).

Now, onto what to build!

10-meter Delta Loop

Our first antenna project is a simple yet effective skywire for getting started in HF activity in high style. It's called the Delta Loop. This popular antenna uses a full wavelength of wire to transmit and receive signals better than a regular dipole. It includes a homebrewed balun, and its overall cost is incredibly low. You get more for less, and "scavenger hunting" for special components isn't necessary. That's good news from any standpoint!

A 10-meter Delta Loop is small enough to fit almost anywhere, plus you can build a larger scale model for terrific DXing results on another band, like 15, 20, or 40 meters. A couple of years ago, I quick-assembled a Delta Loop for 30 meters and installed it beside my 30-meter dipole for comparison. The results were very interesting: in every case, signal reports were better when I used the Delta Loop. Why? It has double the wire (often called signal radiation or signal capture area) of a $1/2$-wave dipole, and its combination of horizontal and vertical polarization is ideal for both near-

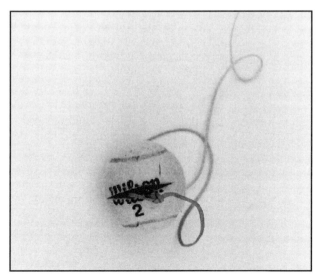

Figure 4-6. An ordinary tennis ball fitted with a fishing weight and tied to a thin nylon rope is an ideal means of installing wire antennas right from terra firma. Just toss the ball over a high tree limb, connect the nylon rope to one end of the antenna's insulator, and pull it up into place.

by and DX contacts. Have I kindled your enthusiasm for building a Delta Loop? Great! Let's get started!

The material you need to build a Delta Loop includes a full wavelength of wire (I used No. 14 insulated wire for my Delta Loop and it works fine), a $1/4$-wavelength piece of 75-ohm RG-59 coax (for the balun), and a length of 50-ohm RG-58 or RG-8X coax long enough to reach between the balun and your in-shack rig. You also need three insulators (two for hanging/supporting the antenna and one for the feedpoint), plenty of lightweight support rope (which ties to each end of the upper insulators), a roll of Coax Seal (for weatherproofing feedpoint and balun/coax connection points), a PL-259 connector with reducer for small coax (for plugging into your rig), and a soldering iron and solder. Although it's not mandatory, you may prefer to go "first class" by adding PL-259 connectors and a double junction connector at the outdoor junction of the balun and 50-ohm coax. If you include these connectors, rather than simply soldering the two cables together, remember to thoroughly weatherproof them as previously discussed. Even the smallest amount of rain can seep into connectors wrapped only with tape and can ruin an otherwise terrific antenna in a very short time. Remember when connecting the two coax cables to wire shield-to-shield and insulate them well with several "wraps" of electrical tape. Then connect the center conductors of the cables and wrap them also with electrical tape to avoid shorts. Double check to ensure the shields and center conductors of coax can't possibly touch each other, then thoroughly weatherproof the whole junction with Coax Seal. (While I'm getting slightly ahead of myself in discussion of this step, I want you to understand how the cables are interconnected so everything is clear in your mind before actual assembly.)

An outline of the Delta Loop is shown in Figure 4-7. Its dimensions are marked in wavelength so you can apply them to any band desired. The amount of wire required for making the loop is determined by the formula 936/Freq. (MHz) = length in feet. For example: 936/28.40 (MHz) = 32.95 (feet). The wire threads through the two top insulators, and each end secures to each side of the bottom insulator. The coaxial balun then connects to the bottom insulator with its shield connecting to one side, and its center conductor connecting to the other side.

The balun is a $1/4$ wavelength of 75-ohm RG-59

Figure 4-7. Outline of the Delta Loop antenna. A full wavelength of wire and coaxial-type balun delivers high performance. This antenna can be scaled for operation on any HF band.

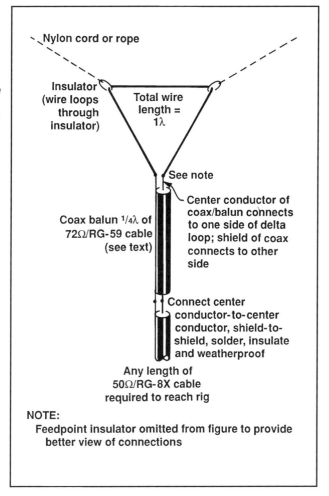

coax cable. Since radio waves travel slower through wire than through free space, we must consider their speed when calculating this $1/4$-wavelength. Looking in a technical manual on coax cable, we find the velocity factor for RG-59 is .66, so the balun's length is .66 x $1/4$ of a wavelength. You can quick-calculate that as 32.95 ÷ 4 = 8.2 x .66 = 5.4 feet (length of 75-ohm coax balun). Alternately, you can use the formula: 234 ÷ Freq. = $1/4$ wavelength (8.2 feet) and multiply that number by the velocity factor (.66) to again derive the balun's length as 5.4 feet. As long as we're having fun with electronic math, let's also calculate the wire's length for 40 meters: 936 ÷ Freq. (7.100 MHz) = 131.8 (feet). Balun length is: 234 ÷ Freq. (7.100 MHz) = 32.9 x .66 (velocity factor) = 21.7 feet (length of balun). One final note: remember to allow 3 to 5 inches for wrapping wire ends and coax pigtail leads around insulators; then lay out, measure and cut each for antenna assembly.

The next step of our project involves cutting the wire to length, attaching it to the feedpoint insulator, and connecting the coax cable/balun. Stretch the wire tightly between two supports, then use a tape measure or yard stick to determine exact length needed. Add 4 or 5 extra inches for wrapping around the insulator, then cut the wire and remove/scrape insulation off each end. Thread the two support insulators through the wire, then attach each end to the bottom insulator. Twist the wire back on itself for stability, then twist the coax shield around one wire and the coax center conductor around the other wire.

You'll then need to separate the coax shield and braid into "pigtails" approximately 3 to 5 inches long. Use a pocket knife to first remove the coax outer rubber jacket, being careful not to nick the shield. If this is your first experience working with coax, practice on an extra piece first. Next, push the shield back toward the main length of coax so it bulges right where you cut off the outer jacket. Now use the tip of your pocket knife's blade or a small pick to separate or make a small opening in the braid through which you can pull the center conductor. Using needlenose pliers, carefully pull or "roll" the center conductor through your prepared hole in the coax shield. Got it? OK, now use your pocket knife again to carefully remove the insulation from the center conductor. Some types of dielectric are quite tough, so be patient and "whittle away" insulation as necessary.

After joining the coax and antenna, use solder and a large soldering iron to secure the connections. When everything cools, protect the junctions and coax openings with a liberal amount of Coax Seal as previously discussed. You're now ready to connect the balun and feedline by pigtailing their coax leads, soldering, taping, and weather sealing. One convenient means of making a neat (balun-to-feedline) connection involves twisting (and soldering!) shields and bending them back in one direction (toward the shack) and connecting/soldering the center conductors while bending them forward (toward the antenna). Once you're sure these connections are well sealed, you're ready to erect the antenna.

Assuming you've selected the tree limbs for supporting each side of the Delta Loop and have installed light support ropes as discussed earlier in this chapter, you're ready to pull your home-assembled antenna into place. Attach the pull-up ropes to each of the support insulators, then sequentially pull the ropes until the antenna is at the desired height and takes the shape of the antenna shown in Figure 4-7. Remember to leave some slack in the pull-up ropes so the antenna can swing in the breeze without breaking. If necessary, lightly tug on the coax cable (balun) so the feedpoint is at the bottom rather than slightly on one side. Now move indoors, install a PL-259 plug on the coax end, connect it to your rig or SWR meter, and congratulate yourself on a job well done!

The Delta Loop is a 50-ohm-impedance antenna and should not require use of an antenna tuner. You should, however, check the SWR to ensure that it's below 2:1 before use. If the SWR is above 1:5 within your desired operating range, make additional SWR checks lower (toward the CW segment of band) and higher (toward the highest end of band). If the SWR is lower at a lower frequency, shortening the antenna's overall (full wave) length by 2 or 3 inches should raise that point of minimum SWR into your desired range. Likewise, if the SWR is lower at a higher frequency, you'll need to add additional wire to the antenna. This situation seldom occurs when the antenna is mounted in the clear and away from other metal objects. If you must add wire, however, do so until the antenna's SWR is ideal at your desired range; but I would strongly suggest replacing the full length of wire with a single and unbroken length.

This antenna radiates maximum energy broadside to the loop, and you can use that effect to maximum

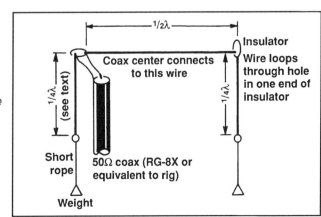

Figure 4-8. Outline of the modified Bobtail antenna. This simple skywire really hauls in the DX and is easily assembled for any HF band.

advantage by your choice of supports. If you live in California, for example, antenna supports aligning the loop's ends north and south will give you maximum radiation to the east (most of U.S.), and to the west (Pacific areas). Another helpful tip: if you assemble a Delta Loop for 10 meters, you may prefer to assemble it "rigid style," that is, using all-aluminum tubing rather than wire. In this case, use your ingenuity to add non-metallic/wood supports and maybe a hinged-type base structure so the antenna can be rotated 90 degrees for directivity.

Here's wishing you good DXing success with the Delta Loop. It's low in cost and high in performance!

Modified Bobtail Antenna Packs a Big DX Punch

This incredibly simple antenna could easily be described as a "sleeper" because it consists of little more than a hank of wire and length of coax, yet it works DX like a champ. The design is a variation of the famous Bobtail Array, it doesn't require any type of balun, and uses a full wavelength of wire to pump out a big signal. Radiation is broadside to the wire(s) and low on your distant horizon, which means it's better for long-haul DX contacts than in-country QSOs. If you're looking for an antenna with a good bang for the buck, this one is a winner! I quick-assembled a 20-meter version of this gem while visiting the Gulf Coast a few years ago and positioned the wires northwest by southeast so its maximum (broadside) radiation was directed toward Europe and Australia. My DX results were amazing! That statement really takes on added meaning when you realize the antenna was swung between corners of a sundeck, and its highest point above ground was only 20 feet. What else can I say except try one yourself—you'll love it.

A sketch of the modified Bobtail antenna is shown in Figure 4-8, and the dimensions are marked in wavelength so you can scale it to any desired band of operation. Length of the vertical wires is determined by the formula $234 \div$ Freq. = length in feet. For example, the calculation for 15 meters would be $234 \div 21.10 = 11.09$ feet. Length of the horizontal $1/2$-wave wire is determined by the formula $468 \div$ Freq. = length in feet. For example: $434 \div 21.10 = 22.18$ feet.

Notice in Figure 4-8 that the "right" wire is actually

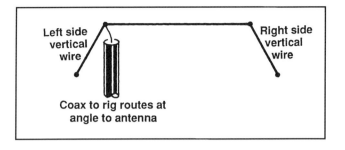

Figure 4-9. Top view of modified Bobtail antenna illustrating orientation of coax feedline.

a single, unbroken conductor positioned horizontally for a $1/2$ wavelength and then vertically for $1/4$ wavelength. For ease in measuring and cutting, you can add your two previous calculations and then loop the wire through an insulator at the appropriate point. For example: $11.09 + 22.18 = 33.27$ feet overall length, with "corner insulator" at 11.09 feet. This antenna really is a cinch to build! All set now to head out to your yard with wire, cable, and tie-up rope? Good!

After measuring and cutting the antenna's two wires, you're ready to install the corner insulator and attach the coax cable. Remember to add the 3 to 5 extra inches for wrapping the wire back on itself as we discussed above. Then add the corner insulator and connect the coax shield to the $1/4$-wave (shorter) wire and the coax center conductor to the longer $1/2$-plus-$1/4$-wave wire. Apply Coax Seal to weatherproof the coax opening/end where you separated pigtail leads. Next, add end insulators and medium-size weights so the antenna will assume its proper shape when raised into position. Finally, attach pull-up ropes to each side and raise the antenna into its final position. You're now ready to finish your installation by routing the coax into the house, installing a PL-259 connector to mate with your rig, and checking SWR. All that's left is to make some good DX contacts!

Let's now discuss some additional notes and tips to ensure "first time" success with the modified Bobtail antenna. First, try to erect it in the configuration shown in Figure 4-8. This antenna will work fine even if the ends of its vertical wires are only a few feet above ground; but, as with any antenna, raising it above near-field obstructions always improves performance. If buildings, autos, chainlink fences, etc. are within a $1/4$ wavelength of the vertical wires, you may experience interaction in the form of an SWR above 1.5:1. If this occurs, use your transceiver's built-in automatic antenna tuner or a manually operated anten-

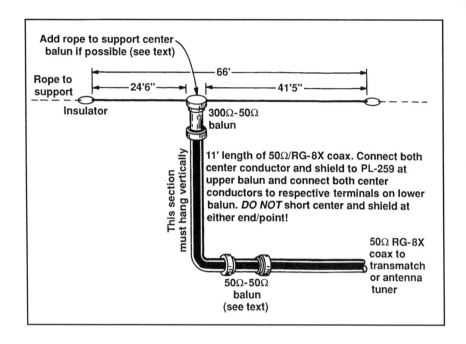

Figure 4-10. Assembly outline of the popular Carolina Windom antenna. A 4:1 balun is used at the feedpoint and an in-line 1:1 balun is used to isolate its vertical radiating section from the main transmission line. The antenna covers all HF bands from 80 through 10 meters, including WARCs.

na tuner to achieve a near-1:1 SWR. Finally, try to route the coax away from the antenna as illustrated in Figure 4-9. Avoid routing the coax directly beside a vertical element. If you're hanging the antenna from an upper floor of an apartment complex, the coax can be routed horizontally and at right angles from the antenna directly into your shack. Final "pruning-to-frequency," if desired, can be accomplished by trimming each end of the antenna while monitoring SWR.

I can't emphasize how well this antenna works despite its simplicity. It's a real little tiger!

The Carolina Windom: A Hot Multiband Skywire!

Let's say you earned that highly sought license for HF operation, secured a new transceiver (or at least new-to-you), and you need a good multiband skywire for working the world. The Carolina Windom is the perfect answer! This little delight works like gangbusters on the Novice/Tech bands of 10, 15, and 40 meters, plus it will also radiate your signal on 30, 20, 17, and 12 meters after you upgrade to General class. That's right: a seven band wire antenna that thinks it's a beam...and you'll think so too after a few contacts. Do I jest? No way, friends. In fact, the DXing results

I got were so close to those from my three-element triband beam, I even erected a reference dipole to determine if my beam had lost its "oomph." The results were quite interesting: the beam still had its originally measured gain over the reference dipole, but the Carolina Windom's performance was more akin to the beam than the dipole!

There is one minor hitch, however: you have to use an antenna tuner with the Carolina Windom to reduce its SWR below 1.5:1 on most bands. The high SWR won't affect this antenna's performance, but it will cause your transceiver to reduce power output. Any basic (manual) coax-to-coax tuner or the automatic antenna tuner built into most deluxe HF transceivers usually works fine with the Carolina Windom. The automatic-type tuners are really neat, too, as you simply install the antenna, connect the transmission line to your transceiver, press one button, and are ready for operation. So, how do you assemble this big-time antenna? Read on!

An outline of the Carolina Windom is shown in Figure 4-10, and you'll notice that this antenna is unique in several respects. Its two horizontal sides are unequal in length, two baluns of different types are used, and coax cable is separated into two sections. The horizontal wire(s) may be mounted sloping or angled to mate with available supports, but the 11-foot coax section between the two baluns operates as an integral part of the antenna and should always be positioned vertically. A popular type of balun, known as 4:1, can be installed at the feedpoint, and any type of 50-to-50-ohm (1:1) balun can be installed between the two sections of coax cable. Since you'll probably be using a "barefoot" transceiver with 100 watts output, I suggest using small and lightweight baluns. An SO-239 socket on the 4:1 balun will make it easier to connect the 11-foot piece of coax with a PL-259, but you'll probably need to connect coax "pigtail" leads to the top end of the 1:1 balun. Just remember to keep wire lengths short, wrap them with electrical tape to prevent short-circuiting, then weatherproof both junctions with Coax Seal. You can then plug in an extension-transmission line cable with PL-259s on each end between the balun and your transceiver. While "dinking" with the Carolina Windom, I found a transmission line length of approximately 85 feet resulted in a very low SWR on 40 meters and 10 meters, and use of the automatic tuner in my transceiver was not required

71

on those bands (but was necessary on other bands). I don't recommend using 85 feet of cable in every case, however, as it may encourage you to coil up the excess in a corner rather than cutting it off, or it may not produce the same low SWR in your case due to installation variations. Since two baluns tug at this antenna's feedpoint, it should be supported by a lightweight rope. I would suggest using rope at the feedpoint and at each end.

This antenna works best when it's installed high and in the clear. Don't fret, though, if you don't have a high installation point; it's more important for the antenna to have a "clear view" for at least 20 feet from the vertical coax section between the two baluns. Start collecting parts and wire to build a Carolina Windom soon. I'm sure you'll like its performance and its ability to switch bands in a flash!

The "El Toro" Special

Many years ago, a simple yet impressive antenna known as the El Toro was popular among amateurs with limited funds and high enthusiasm. Its popularity faded over time and other "trendy" types took its place, but the El Toro works just as well today as it did during the 1950s. Let's revisit this easy-to-assemble skywire.

Simply explained, the El Toro is a $1/4$-wavelength antenna built using readily available 450-ohm, open-air feedline, which is also known as ladder line (see Figure 4-11). Ladder line is quite inexpensive and available from amateur dealers nationwide. The antenna can be erected as a vertical, a sloping wire, or bent to mate with various supports/surroundings. It's RF-fed with regular 50-ohm coax cable, and, since one "side" of the antenna is connected to the coax shield, the usual ground losses associated with $1/4$-wave antennas are minimized. In other words, the El Toro usually works as well as a ground-mounted vertical with 120 radials (whew!), or better than a ground-mounted vertical with only six or eight radials.

There is one hitch, however. You do need a solid ground connection at the base/coax feedpoint of the El Toro. The easiest way to secure this ground connection involves positioning the antenna so its feedpoint is near your home's (outdoor) cold water pipe. Use your pocket knife to scrape a 1- or 2-inch section of

Figure 4-11. Outline of the simple but effective "El Toro" antenna. This classic radiator delivers surprisingly good results at very low cost.

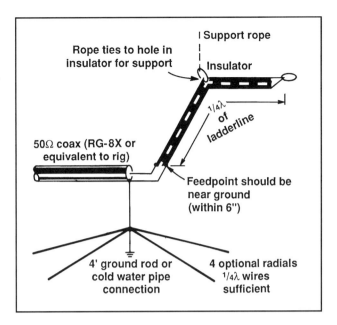

the pipe till it's shiny clean, then attach and clamp your antenna's ground connection to the pipe. Remember to weatherproof this connection well! One additional note: due to this antenna's short length and the probable location of your home's outdoor cold water pipe, it's usually best suited for lower frequency bands like 40 or 80 meters. Here's your answer to getting some good CW practice on the lower Novice/CW bands and working nearby states needed for your Worked All States (WAS) award.

The materials you need to assemble an El Toro antenna are a $1/4$-wavelength section of 450-ohm ladder line and length of 50-ohm cable to connect between it and your transceiver. The antenna's length is calculated using the formula: 234 ÷ Freq. = length in feet. For example, 40 meters is calculated as follows: 234 ÷ 7.100 = 32.95 feet; 80 meters is calculated as 234 ÷ 3.60 = 65.0 feet. Unroll the ladder line, being careful to avoid twists and catches, measure the appropriate length, and add a couple of extra inches for twisting the "far end" together (remember to solder that point!) and for attaching coax at the feedpoint. If you plan to erect this antenna in a bent fashion, estimate approximately where the bend should be positioned, and wrap small nylon rope around one conductor at that point to hold it in position. Next, connect the feedline/50-ohm coax cable to one conductor of the ladder line and connect the shield to the other conductor (which also connects to your cold water pipe/ground wire). Double check that the coax center conductor and shield aren't shorted, weatherproof all connections, toss pull-up ropes over tree limbs as discussed earlier in this chapter, and raise the El Toro into position. With everything in place, the pull-up ropes can be used to stabilize the main section while small pieces of rope steady the base section.

The check-out procedure and operation of the El Toro should pose few problems, as this antenna is quite easy to "get going." But if you experience high SWR, it's probably due to nearby objects affecting radiation. You can prune 1 or 2 inches from the antenna at its base (you did leave some slack for wind sway, didn't you?). Alternately, just switch on your rig's automatic antenna tuner and let it handle frequency-tweaking for you!

Now get cracking on some good late night CW QSOs and experience the fun of hamming after 10 meters "closes" and the local 2-meter crowd is fast asleep!

The Bruce Array: A High-Powered Antenna!

Almost every new amateur dreams of owning a big beam antenna mounted on a tall tower, but cost considerations involved usually mean we have to start with much less expensive skywires. Wouldn't it be nice if a wire antenna strung in the yard gave you the big signal gain you'd get from a beam? Well friends, the Bruce Array fills that bill. It's a simplified form of the famous "Curtain" antenna used by shortwave stations, such as the Voice of America, and it works like a bandit on 10 meters and other upper bands. The Bruce Array could be scaled to work another band, but you'd probably need a large farm to support it! This antenna is a bit challenging to erect and maintain through storms, but it's a real DX grabber! Maximum signal radiation is broadside to the wires, so your first step in planning this gem is to visualize its size and pick out tree limbs to support it in appropriate positions. Some zig zagging or bending will probably be necessary; just accept as few compromises as possible and go for the gusto!

An outline of the Bruce Array is shown in Figure 4-12. It's built out of regular antenna wire plus 300-

ohm TV "twin lead" for phasing sections. A regular 300-ohm-to-50-ohm balun is connected at the feedpoint, then 50-ohm coax is directed to your indoor transceiver. Notice each of the horizontal sections closest to the feedpoint is a $1/2$ wave long by a $1/4$ wave tall. In the case of 10 meters, these dimensions are calculated as follows: 468 ÷ 28.4 = 16.47 (long) and 234 ÷ 28.4 = 8.2 feet tall. The end sections are a $1/4$ wave square, or 8.2 feet.

You can add additional end sections to really "hop up" performance and make a super antenna, but remember, added sections must "balance." In other words, be sure to add one on the left if you add one on the right. You can also continue expanding this antenna until you run out of wire or space! If you expand the antenna, replace the single vertical end wire (on each side) with 300-ohm TV twin lead twisted one-half turn as in the middle sections (again, refer to Figure 4-12). I've heard of only one amateur who assembled this antenna with multiple sections. He added six ($1/4$ wave square) additional sections to each side, strung it through trees in woods near his rural home, and worked DX on 10 meters when others couldn't hear signals and thought the band was closed! Needless to say, this is a big-time project with equally big-time results!

When selecting 300-ohm TV twin lead for the phasing sections, try to get the heavy duty type with large center conductors. This twin lead is better quality, less prone to breakage, and handles higher RF power levels. Standard 300-ohm twin lead is suitable for power

Figure 4-12. Outline of the famous Bruce Array. This super skywire radiates a whopping signal.

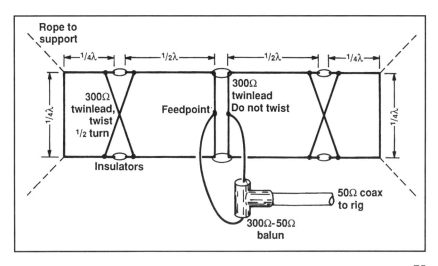

levels up to 100 watts, however. Notice that the twin lead is twisted one-half turn between each of the antenna's sections and is connected only at the top and bottom, not in the middle. You should lay the twin lead flat and maybe mark one side with fingernail polish to ensure that you twist it only one-half turn during assembly. Also notice that the center piece of 300-ohm TV twin lead is not twisted and is broken in the middle for connection to the balun. An easy way to make this splice involves measuring two pieces of twin lead, 4.1 feet each, then connecting them to the balun and to the ends of the center wires as shown in Figure 4-12. Be patient when assembling this monster: twist all wires tightly to form a solid structure, solder them securely, and weatherproof everything while it's on the ground!

Getting ready to erect the Bruce Array into position will probably be an all-day process, so work patiently and have plenty of nylon pull-up rope on hand. After tossing your pull-ropes over your selected tree limbs, as discussed earlier in this chapter, you can move between the various ropes to raise the antenna into position. Guiding ropes will be extremely helpful at this time! Ideally, the antenna's lower wires will be 20 or 25 feet above ground, but you can also get good results even with the bottom wires only 10 feet above ground. Less ground clearance invites curious hands, however, and should be avoided.

The Bruce Array is the "King of the Wire Antennas," and it will put you on top of the crowd as long as it stays up. Enjoy!

"Invisible" and "Secret" Antennas

Many amateurs live in apartments or in environmentally sensitive areas and are restricted to using undetectable antennas, yet they achieve very good hamming results. What's the secret to their success? Do they use some sort of small magical antennas? Not really: their antennas are the same types as described earlier in this chapter, but are made with smaller gauge wire and coax. Most "invisible" or "secret" antennas fall into this category. Invisible antennas are usually the result of personal ingenuity and may take any form, such as a long plastic pipe supporting a flag in front of a house (the proverbial flag pole antenna), an attic-mounted antenna, indoor vertical disguised as a pole lamp, or even a small beam painted to blend with its surroundings.

All of our previously described antenna projects can also be assembled as invisible antennas by using fine wire, small insulators, and miniature coax cable. Additionally, wire can be dipped in blue/gray paint so it practically disappears against the sky when erected into place, and coax cable can be painted brown, gray, green, etc. to match structures in its line of sight. A good choice of wire for invisible antennas is No. 24, but even smaller gauge (right down to a size resembling spider webs) can be used when it's well supported. Ultra small 50-ohm coax, designated RG-174/U, is readily available from dealers nationwide. This cable can be used in lengths of up to 100 feet and power levels of up to 250 watts without problems. Since you're assembling an exceptionally lightweight antenna, insulators can also be homebrewed from small objects like cut thread spools, plastic ties, or even nylon rope. Just think small!

Are you getting the idea that invisible wire antennas are fragile and prone to breakage? That is true, so support your prize antenna in several places and always be prepared to replace it after storms. You have an extra advantage in that each new invisible antenna erected can be better than the last one. To help keep your outdoor work from being discovered by suspicious eyes, early evening hours (right before dark) may be an optimum time for installing your antenna. And, of course, preplanning is very important so everything goes together without attracting attention.

Earlier I explained that invisible or secret antennas can take many forms, such as a multiband vertical mounted inside a plastic pipe used as a flag pole. If you choose that route, a cut section of garden hose lying between the flag pole and the house shelters its coax cable inside. Other secret antenna ideas include a regular dipole or Carolina Windom lying on a roof rather than appearing obtrusively above it, or even a full-size vertical antenna mounted on a rotor so it may be tilted flat (and unseen) against the roof or raised into position for operating. Finally, one of my own "undercover operations" may inspire your creative thinking. A few years ago, I lived in a large apartment complex and was faced with the problem of antenna installation. I studied the roofline and found that a 4-foot high facade surrounded each building to hide air conditioning units mounted on the flat roof. A brief discussion with the management convinced them that I could safely install an antenna on the roof that would not interfere with television reception and would not be visible from the ground. During a weekday, when most people were away, the management even loaned me a ladder to climb on the roof and install my antennas. I strung a Carolina Windom (disguised with blue-gray paint) between inside sections of the facade and across most of the roof, added a 30-meter dipole diagonally across the roof, and, believe it or not, hauled up a 4-foot roof tower and three-element triband beam. After mounting the beam and rotor on the roof tower, I painted it with non-metallic paint to match the sky. The coax cables were then painted to match the building, and dropped into the shack. After all antennas were installed, both tenants and management could look directly at my townhouse and never see an antenna! Inside, I worked all bands in high style! Sometimes the most often overlooked "disguises" are natural structures!

We hope you enjoyed the antenna projects presented in this chapter, and will assemble at least one in the near future. A fascinating world of HF communications awaits you, and I heartily encourage you to join the fun as soon as possible. Good luck and good DXing!

CHAPTER 5
Special Treats for VHF Enthusiasts

As we all know, 2-meter and 70-centimeter activities are super-hot areas of interest among newcomers and seasoned amateurs alike. In fact, approximately 85 percent of all presently active licensees own at least one VHF or dualband FM transceiver that they use for casual chats and/or emergency monitoring. Such popularity naturally warrants recognition, so this chapter describes a variety of neat first-time projects that are low cost and easy to duplicate. I'll begin by outlining some simple items for mobiling and portable hamming, then conclude with some ideas and projects to kindle your interest in OSCAR satellites. If you've been looking for an ultra low-cost mobile antenna and mount, a simple noise filter for the mobile rig, or maybe a battery-level monitor for an FM handheld, this chapter has the answer!

Many newcomers visualize OSCAR satellite com-

Figure 5-1. Here's an inexpensive and easy-to-assemble $1/4$-wave 2-meter mobile antenna and mount. This homebrewed gem is shown installed on W4CEC's vehicle.

munication as a complex and expensive pursuit, but that is definitely not the case. At least one low orbiting Russian amateur satellite presently awaits your investigation, and you can contact other amateurs through this satellite while transmitting or "uplinking" with your 2-meter rig and receiving or "downlinking" with a 10-meter receiver. Furthermore, large antenna systems and high power equipment are not necessary for using these "RS" (Radio Sputnik) satellites. Some amateurs achieve good results using only a 2-meter handheld transceiver, connected to a vertical antenna for transmitting, and a simple 10-meter dipole connected to their HF transceiver for receiving. This promises to be a very exciting chapter, so read on as we jump right into some fun-filled projects.

Mobile VHF/UHF Antenna and Mount for Trunk or Hood Installation

Looking for an inexpensive, but effective, custom antenna and mount that blends in with your vehicle's lines and color? Check out our trim little assembly illustrated in Figures 5-1 and 5-2. This antenna and mount attach to the inner lip of your trunk or hood via sheet metal screws, the bracket can be painted to match your vehicle, and it results in an almost factory-quality appearance.

Just raise your vehicle's trunk lid and notice the convenient inner lip on either side. The bracket attaches to that lip, extends upward, then "folds over" the fender so it clears the trunk lid when it's opened or closed. If you prefer a hood-mount, the bracket is bolted to the fender's inside lip and its bottom bend is disregarded. Trunk lips and fender wells on various vehicles vary in size, so apply your own measurements to the following "generic" description.

The base/mount section is made from a piece of $1/8$-inch thick aluminum strap, approximately $1\text{-}1/2$ inches wide by 4 inches long. The strap is bent at right angles so the bottom section is $1\text{-}1/2$ inches, the middle section is $1\text{-}1/2$ inches, and the top section is 1 inch. The strap can be quick-formed using a bench vise and hammer, or a local heating and air conditioning company might cut and bend the bracket for you at a low cost. Drill two small holes in the lower section for passing sheet metal screws (which will attach to the trunk or hood's inside lip). Then drill a $5/8$-inch-

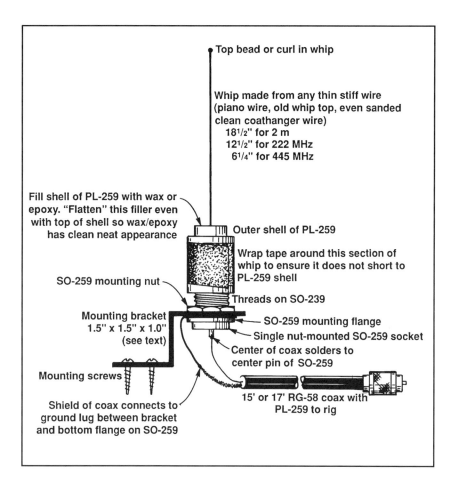

Figure 5-2. Assembly outline of the simple $1/4$-wave antenna. This antenna can be configured for 2-meter, 222-MHz, or 70-centimeter operation.

diameter hole in the bracket's upper section and install an SO-239 antenna socket. This socket is available in two styles: one mounts via four small screws, and the other (which is easier to install) mounts by a single nut threaded over the SO-239.

Finished? Great! Now connect a 15- or 17-foot length of RG-58 or RG-8X 50-ohm coax cable to the SO-239's terminals (center conductor to center of socket and shield to outer/ground terminal). Use a high wattage soldering iron to ensure a good solid connection. Double check your work, weatherproof all exposed areas with Coax Seal, then set the mount aside.

Let's now assemble the antenna that will plug into your mount. Although you could go fancy with an open-air center loading coil for dualband operation, let's keep things simple by building a basic $1/4$-wave whip. Approximate length of this whip, as measured from the top end of the PL-259 plug to the tip of the whip, will be 18-$1/2$ inches for 2 meters, 12-$1/4$ inch-

*Figure 5-2A.
An exploded view
and assembly
outline of
Figure 5-2.*

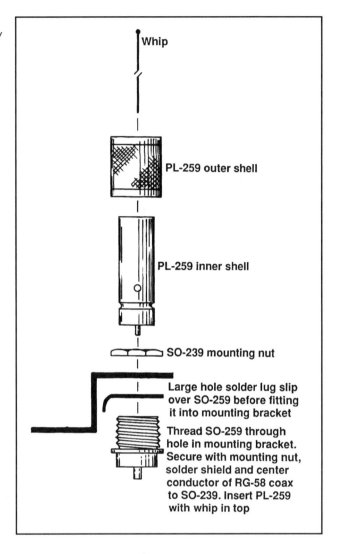

es for 222 MHz, or 6-^{1}/4 inches for 445 MHz. Generic whip material is usually available from amateur dealers nationwide, or you can salvage a whip from the top section of a discarded HF mobile antenna or even use piano wire. The latter is a good choice because it's stiff, flexible, and quite thin.

Push the whip through the PL-259's center pin, crimp as necessary for a firm fit, and solder the whip to the PL-259's center pin. Be sure the whip remains straight and does not touch the PL-259's sides! If in doubt, wrap a few turns of plastic electrical tape around the whip's lower area (the area that centers inside the PL-259's body). If desired, a small coax-type adapter may now be threaded into the PL-259.

Next, while making sure the whip is straight, fill the PL-259's inner area with wax or epoxy to hold everything secure and to provide insulation and weather protection. After final check-out and length pruning, you can twist a small curl at the whip's top or add a small bead. Additionally, extra whips made for other bands can be swapped quickly and easily on the completed mount. That's right: three interchangeable antennas for less than $5! The whips can be painted black or a color to match your auto, just remember to use nonmetallic paint like Krylon. I used this antenna and mount idea for several years and loved it. May you also find it a low-cost delight!

Voltage Spike and Polarity Protection for Mobile Transceivers

Both VHF and HF transceivers used in autos are often subjected to accidental power lead reversal and dangerous inductive voltage spikes caused when starting the auto with the rig in operation. Luckily, there's an easy solution to these problems!

Referring to Figure 5-3, you see that a power diode capable of handling the maximum current required for a particular transceiver (10 amps is usually fine for FM transceivers up to 50 watts output) is wired in series with the rig's positive battery lead. The diode prevents opposite-direction current flow when connecting cables under the dash at night. The capacitors you see in Figure 5-3 bypass voltage spikes to ground and protect the transceiver from surges during starting, air condi-

Figure 5-3. Circuit diagram of simple voltage spike and polarity protector for mobile rigs. Its capacitor values aren't critical and "hamfest specials" can be used successfully.

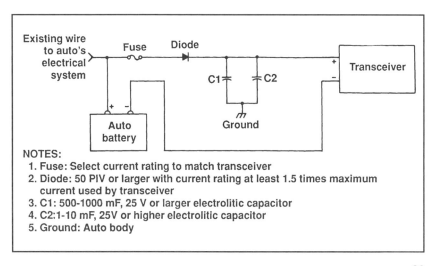

NOTES:
1. Fuse: Select current rating to match transceiver
2. Diode: 50 PIV or larger with current rating at least 1.5 times maximum current used by transceiver
3. C1: 500-1000 mF, 25 V or larger electrolitic capacitor
4. C2: 1-10 mF, 25V or higher electrolitic capacitor
5. Ground: Auto body

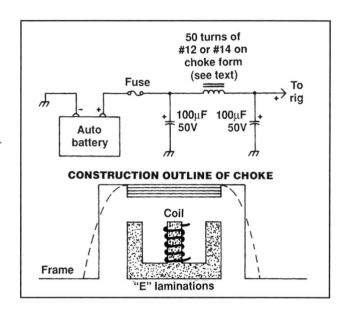

Figure 5-4. Homebrewed alternator filter for mobile transceivers. The choke is assembled by winding a new coil on a discarded power supply's choke core.

tioning switching, etc. The value of these capacitors isn't critical and may range from 10 mfd to 1000 mfd @ 50 volts or higher. The diode and capacitors are often available at low cost from hamfest fleamarkets.

Installation and placement of the diode and capacitors will be determined by your auto's available space, type of rig, and power cable used and will vary with each installation. Simply remember to heat sink the diode as necessary and weatherproof the enclosure if it's mounted in an exposed area. This project may seem overly simple, but it could easily "save the day" by minimizing your future repair bills.

Heavy-Duty Alternator Noise Filter

During contacts with mobile operators, you'll occasionally hear a high-pitched noise or "hash" on their transmitted signal. This whine varies in pitch according to the motor's speed and is easily removed with a simple filter installed in the transceiver's positive power cable lead. Low power filters, such as those made for 5-watt CB sets, will eliminate the whine, but their current handling abilities are usually inadequate for higher power amateur transceivers. Where do you find an appropriate filter? Homebrew it, naturally! A diagram and outline of such a "big-time" filter is shown in Figure 5-4. It's connected in series with the transceiver's positive battery lead and may be conveniently mounted under the auto's dash, or even under the hood if you prefer.

The filter is assembled using a discarded power supply choke's metal core, approximately 3 x 3 x 1 inches (HWD). A "good" choke isn't necessary, as you'll completely disassemble it and wind your own new inner coil. First, using pliers, you'll remove the choke's mounting bracket, then locate the bar on top of the choke's "E" lamination and tap it loose with a small hammer and screwdriver. Discard the choke's original winding (found inside the "E") and replace it with your own coil, consisting of approximately 50 turns of No. 12 or 14 insulated wire. Then replace the choke's top bar and mounting bracket, wrapping the completed choke with electrical tape. Again, capacitor values for this filter aren't critical, and any value between 100 and 1000 mfd will work fine. Connect the capacitors to wires on each end of the choke, making sure that their positive terminals connect to the choke's wires and their negative terminals connect to the transceiver's negative power lead. Solder all connections, tape everything securely to avoid short circuits and blown fuses, then mount the filter in an appropriate location.

This homebrewed filter does a very good job of eliminating mobile noise on your transmitted signal. It can be assembled in a very short time, and almost any size choke capable of housing your "scrambled wound" coil works fine. Avoid using wire smaller than No. 14 gauge because it must pass all the current required for your transceiver. Now hit the open road in style and enjoy a nice clean transmitted signal!

Quick and Easy Battery Monitor for FM Handhelds

Handheld transceivers are terrific rigs and great traveling companions, but their battery packs always seem to reach depletion at the most undesirable times. That's especially true if you have a new handheld or fail to keep track of the number of hours you use between battery recharges. Some FM handhelds feature a built-in battery monitoring circuit or alarm, but they typically indicate when the battery pack becomes "dead" (which is obvious without an alarm!) rather than when the pack is approaching that stage. Frequently "topping up" a battery pack so it doesn't

Figure 5-5. Circuit diagram of the battery monitor circuit for FM handhelds. It's simple but very effective in forewarning you of an upcoming discharge condition.

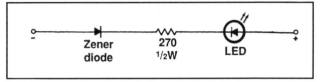

fail at an inopportune time is not a good alternative, as nickel cadium cells (which power most handhelds) develop a "memory" and soon drop to full discharge when reaching their customary "topped up" point. A perplexing situation? Not really: whip together the simple battery monitoring circuit shown in Figure 5-5 and you'll enjoy more dependable battery-powered operations and ensure a long full-cycle battery life. That's a two-in-one advantage you'll love!

This monitor's operation takes advantage of the fact that there's a slight drop in a rechargeable battery pack's output voltage right before it requires recharging, and that sign of approaching discharge can be detected by a zener diode. You then have time to reduce output and revert to shorter transmission times for continued operating without fading into the noise due to dead batteries.

Referring to Figure 5-5, you see that a $1/2$-watt resistor, LED, and low wattage zener diode (with a voltage rating between the battery pack's full charge and discharge level) are connected in series to produce this monitor circuit. The zener conducts in its reverse, or avalanche, direction until the battery voltage drops below the zener point. The LED (which was previously illuminated) then extinguishes, warning the operator of a weak battery. The zener, LED, and resistor form a series circuit, so they can be assembled in any sequence desired. I built several of these monitors for my handhelds and QRP transceivers, one of which is shown in Figure 5-6. In my design, the monitor plugs into the rig's DC power socket, but you may prefer to assemble it with regular screws as positive and negative terminals that can be touched to the bottom of your rig's battery pack for impromptu testing. This "hold-to-check" monitor design is perfect because you really only need to check battery condition occasionally.

Exact values for the resistor and zener depend on your battery pack's voltage. For instance, to check a 12- or 12.5-volt battery pack, combine a zener in the 10.8- to 11.5-volt range with a resistor of approximately 270 ohms and an LED of your choice. For 7.2-

to 7.6-volt battery packs, combine a zener of between 6.5 and 7.0 volts with a 270-ohm resistor and LED of your choice. After assembling the battery monitor, you can calibrate it with the aid of a variable power supply and voltohmmeter (VOM). It is possible to calibrate the monitor with only a battery pack, but you'll need to continuously check the voltage as the batteries discharge, which can be time consuming. Or you can use your VOM to check the voltage of a recently charged battery pack, then connect your meter across the variable-output power supply, and set voltage equal to that measured level (simulating full charge). Your assembled monitor can then be connected across the power supply and its operation checked while reducing supply voltage. The LED should stay lit until the pack's "knee of discharge curve" is reached. It should then extinguish within a .1-volt range, say between 11.6 and 11.7 volts with a 12-volt battery pack (that usually measures 12.5 volts with full charge) or 6.8 to 6.9 volts with a 7.2-volt pack (that usually measures 7.5 or 7.6 volts when fully charged). The LED's brightness and the zener's point of impending-discharge condition can be varied slightly in either direction by changing resistor value. When the proper resistor value for a particular pack is found, the LED will be lit until battery voltage begins to drop, then extinguish, and warn of upcoming discharge.

Using this monitor is simple and easy. Just hold the monitor to your handheld batteries' charging terminals and check the LED while you're transmitting at the highest power level. The LED should burn bright-

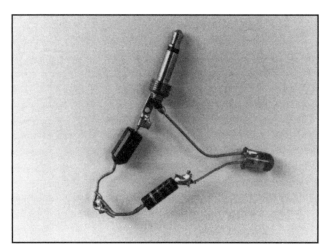

Figure 5-6. Here's a picture of the battery monitor circuit I built for use with a low-power transceiver. The components may be wired "open air" style, but it works great!

Figure 5-7. The quick and easy mobile charger assembled and ready for use. A 71-ohm resistor is mounted inside the auto cigarette lighter plug, and the coaxial plug on the end of the cable mates with the handheld's battery charging circuit.

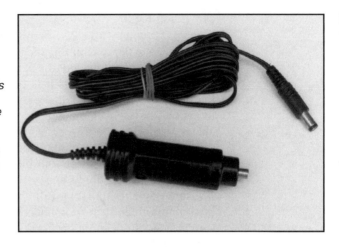

ly if the battery pack is adequately charged. When battery charge drops, the LED will extinguish. You'll also notice when switching to low power that the LED will once again light, indicating that additional transmit time is available. Once you become accustomed to this little gem, you'll never leave home without it. The monitor is simply terrific and costs only pennies to assemble. Go for it!

Quick and Easy Mobile Charger for Handhelds

Although every FM handheld transceiver is factory-supplied with a basic "wall charger" for home use, there are numerous times when a simple mobile charger is also helpful. If you have a spare battery, for example, it can power the handheld while the other pack is being charged in your vehicle. As one pack becomes discharged, it can then be exchanged with the charged pack to give you almost continuous operating ability. A mobile charger is also an invaluable asset during emergencies and AC power outages.

Since a vehicle's battery can supply a large amount of current, the handheld battery pack must be protected from extreme overcharging or it can be destroyed. The simplest and easiest means of providing this protection involves using a resistor connected in series with the vehicle's battery and the handheld's battery pack. The exact value of this dropping resistor is found experimentally by monitoring battery pack charging current while inserting various "test resistors," but its approximate value will probably be between 18 and 95 ohms. Figures 5-7, 5-8, and 5-9 provide details for building this mobile charger.

As a rule of thumb, a handheld battery pack can be charged at $1/10$th its ampere hour (Ah) rating for 10 hours (plus one or two extra hours to overcome natural charging losses). A 600-mAh battery pack can thus be fully charged from "flat," at 60 mA, for 11 to 12 hours. A charging rate of 100 mA will reduce charging time to approximately six and a half hours, and a fast charge rate of 160 mA will reduce charging time to approximately four hours. This "rapid charge" rate is often marked on the side of stand-alone rechargeable batteries. Also, the charging voltage must be slightly higher than a battery pack's rated voltage to "push" current into the pack and recharge it. Connecting a 72-ohm, 1-watt resistor in series with the vehicle battery and handheld's battery pack will usually provide a 60-mA charge current to a 7.2-volt rechargeable battery pack from a 12.5-volt vehicle battery (12.5 - 8.2 = 4.3 volt drop ÷ .06 amp = 71 ohms). When the engine is running, the voltage will rise to approximately 13.0 volts or more, increasing the voltage drop and the potential current flow. Can a 12-volt battery pack be charged from a vehicle's 12.5-volt battery? Yes, but the rate will be extremely low since there's a very small difference in potential (voltage) to push current into the handheld battery pack. If you apply your own calculations for your particular battery pack and vehicle's voltage here, you'll probably find that a resistor of 15 to 18 ohms is required. One final note: remember to avoid transmitting with your handheld while charging it with this basic auto charger. Excessive current passing through the 1-watt resistor will quickly cremate it! Play it safe and just use the charger to "renew" your extra battery pack while traveling.

Figure 5-8. Pictorial diagram for assembling quick and easy mobile charger.

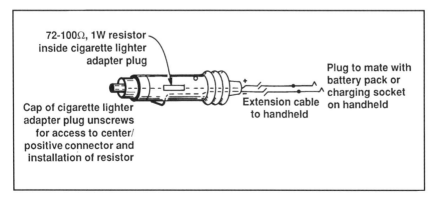

72-100Ω, 1W resistor inside cigarette lighter adapter plug

Cap of cigarette lighter adapter plug unscrews for access to center/ positive connector and installation of resistor

Extension cable to handheld

Plug to mate with battery pack or charging socket on handheld

89

Figure 5-9. Schematic diagram of quick and easy mobile charger. This unit uses only one resistor but works just fine.

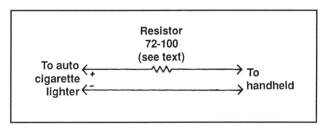

Rather than rely solely on mathematical calculations for your resistor value, I'd strongly encourage you to also use a milliamp meter inserted in series with your test resistor and battery pack, as shown in Figure 5-10. Notice that the milliamp meter's positive terminal connects to the most positive point of this circuit, which is the wire routed to the vehicle battery's positive terminal. The meter's negative terminal then connects to the positive terminal on your handheld's battery pack. After setting up this test arrangement, swap resistor values to vary the current until you reach the milliampere charging rate you want. Finished? OK, let's assemble the charger into usable form rather than just working with open-air wires!

The easiest way to assemble this basic mobile charger involves combining a generic cigarette lighter plug (the type you can unscrew and add an internal resistor to is ideal) with a 6- or 8-foot cable and plug that will mate with your handheld's battery pack (usually a coaxial-type plug). If your battery pack is equipped only with charging terminals, rather than with a socket for a plug, you can fabricate a charging plate (either a simple snap-on version or one that can be held on with a rubberband) using perfboard and screws with double nuts to hold them securely. In other words, just make a custom mount to hold the wires to the battery pack's charging terminals! Use your ingenuity!

Final assembly now involves opening the cigarette lighter plug, unsoldering its center (positive) wire from the plug's tip, and inserting the resistor, the value of which you already determined as described above, in series with the lead. Solder the resistor securely, then wrap its wires with electrical tape to avoid shorts. If you want to double check your work at this point, you can again connect your milliamp meter in series with the charging cable's positive lead and battery pack. Charging current should be the same value as you found during the experimental process.

Lastly, calculate the approximate charging time for a fully discharged battery pack, write it on a small piece of paper, and tape it to the cigarette lighter plug/charger cable for future reference. Who knows: maybe you'll even make three or four charging cables for different battery packs and handheld transceivers! Good luck and enjoy continuous handheld transceiver operation without worry of depleted battery packs!

Triple Rate Charger for Home Use

Occasionally, a new amateur secures a used handheld that comes with only its standard wall charger. He or she might like a more elaborate/fast charger but can't afford a commercially available unit (if one is still manufactured for the rig!). The charger described here is designed to fill that need. It includes many features found in deluxe "drop in" or rapid chargers, but makes liberal use of "junkbox" or hamfest special components to keep the cost low. Exact assembly techniques can be varied according to your specific wants and needs, plus the enclosure and its connectors can be whatever you have available. If, for example, your battery pack only has charging terminals (rather than a dedicated socket), a clamp-on device with screw contacts mounted on a terminal strip can be used with the charger's main circuitry. If you assemble a more elaborate "drop in" type of charger, you can include the complete circuitry in half of the cabinet and build in a separate shelf (with screw terminals on the bottom) for mating with the battery pack to make a stand-alone item.

This charger's schematic diagram is shown in Figure 5-11. Transformer T1, or one very similar to it, is usually available at RadioShacks nationwide, but any transformer producing approximately 16 to 24 volts at 200 mA should work fine. Capacitor C1 provides filtered DC to the battery pack being charged. Its value is not critical and may be anywhere between 50

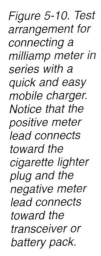

Figure 5-10. Test arrangement for connecting a milliamp meter in series with a quick and easy mobile charger. Notice that the positive meter lead connects toward the cigarette lighter plug and the negative meter lead connects toward the transceiver or battery pack.

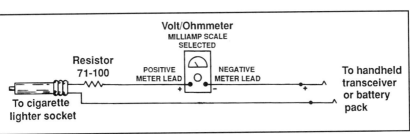

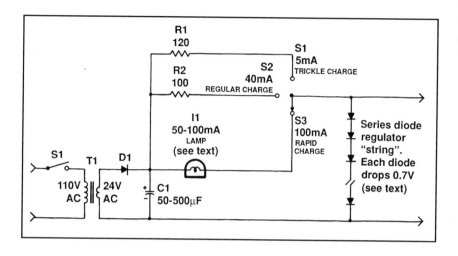

Figure 5-11. Circuit diagram of the triple rate battery charger for home use. It can be assembled in a variety of cabinets or utility boxes, with bottom-mounted screws for charging terminals. Note: never apply high (rapid charge) current to the transceiver (slow charge) socket. Always use battery terminals for that high current.

and 500 mfd, according to what you find available. Resistor values shown for R1 and R2 set the currents indicated for an approximate 12-volt battery pack. (If you want to charge a pack of a different voltage, follow our "experimental value determination" described in the last project and vary the resistance of R1 and R2 accordingly.)

Pilot lamp L1 will set a charging current determined by the lamp's current rating. This is because the lamp is placed in series with the battery pack being charged. The battery pack is capable of drawing much more current, but the lamp serves as a simple and inexpensive regulator. If a 100-mA pilot lamp is used for a 600-mAh battery pack, it limits charging current to 100 mA (and 6.2 hours time will be required to fully charge a depleted pack). If a 225-mA, 7.2-volt battery pack is used, a lamp rated at 25 to 50 mA should be used to limit current. If desired, resistor R2 may be replaced with another pilot lamp of an intermediate current rating, a choice that can be determined by your junkbox or funds.

The series diode regulator "string" connected across the charger's output terminals assures a constant voltage charge and long life for the battery pack. This regulator string is based on the fact that a forward biased PN junction drops .7 volts. The exact number of diodes can thus be calculated by your desired output voltage. This voltage should be equal to, but not more than, 1 volt greater than the battery pack to be charged. Use silicon/power diodes, not glass diodes. Any diode with a rating of 100 to 1000 volts at 500 mA will work fine in the regulator's string.

As previously discussed, how you assemble the charger depends on your own creativity and whatever utility boxes you have available for housing. Remember to check out and calibrate the charger with your milliamp meter to ensure the proper rating at all three switch positions. Your milliamp meter's positive lead connects to the charger's output, and its negative lead connects to the battery pack being charged.

This triple rate charger is used in the same way as any regular "drop-in" charger. You select the rate you want (rapid, regular, or trickle charge), then connect the output terminals to your battery pack. When charging at the fast rate, the pilot lamp will illuminate. Remember to avoid overcharging your batteries for longest life! If, for example, you select 100 mA as a fast charge rate for a (flat/discharged) 600-mAh battery pack, allow only 6.5 hours before removing the pack. The trickle charge rate can be used (left on) a battery pack almost indefinitely without damage. When you later acquire another handheld transceiver, simply recheck charging values and substitute new resistors as necessary, and this basic charger will continue serving you well. Now that's a real asset at low cost!

Key Adapter for Working OSCAR with FM Handhelds

Did my brief mention of OSCAR satellites on the first page of this chapter pique your interest, or did you read right on through it assuming such space-age activities were beyond your budget? Well not so—the secret is operating through low orbit (Phase II-type) satellites like the Russian Sputniks (RS satellites) rather than through complex Phase III satellites like OSCAR 13.

The RS satellites are quite sensitive, easily accessed with low power, and are open for use by amateurs of all nations. Think about that a minute: you can actually communicate with other amateurs via an orbiting Russian satellite, and, assuming you have a 2-meter FM handheld transceiver and 10-meter SSB\CW receiver, additional equipment costs may be less than $25! The main items you need are a key adapter, so you can use Morse code (CW) with your 2-meter FM handheld, and simple high angle-radiating antennas for 2 meters and 10 meters. The key adapter is the subject of this project and is shown in Figure 5-12; the

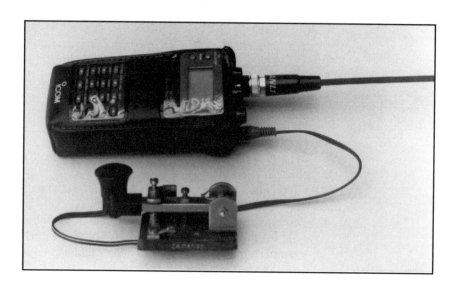

Figure 5-12. Add the simple Morse key adaptor described in this project to your FM handheld transceiver, and your rig can be used for uplinking to the low orbiting Russian satellites. Combine the handheld with a basic 10-meter SSB/CW receiver for a neat, low-cost satellite setup.

antennas will be described in the following project. But before we dive right into the construction details, it's important for you to know some of the basics of satellite operation.

I'll begin by confirming your suspicions that this project involves using CW rather than voice and SSB or FM (FM is absolutely taboo through satellites!). Before you cringe at the thought of operating CW, let me point out the bright side: If you have a home computer and a multimode data controller (like those used for packet radio), the same setup can be used for receiving and sending CW. Yes, you can make many great satellite/CW contacts even if you don't know a single Morse character! The multimode/computer system will decode CW and print it on the monitor's screen for you to read, and you can reply (in Morse code) by simply typing messages on your computer's keyboard. What could be easier? And there's a good possibility it will also improve your Morse proficiency for license upgrading. Now let's talk more about satellites as multi-signal relaying stations in orbit.

RS-10 is one of two wideband transponders aboard the Russian satellite COSMOS 1861. The other transponder is called RS-11 and is a "back-up" should RS-10 develop internal problems. RS-10 also has a few close cousins, RS-12, RS-13, and RS-15, plus a recently launched new brother, RS-16 (not yet fully operational at press time). These satellites are all equipped with 2 meter-to-10 meter capabilities (called Mode "A") that can be switched into operation by ground control stations in Russia. While some of these satellites have additional capabilites, amateurs may still be assured that 2 meter-to-10 meter satellite operations will continue for several years, and possibly much longer.

Transponders such as those aboard RS-10/11, RS-12/13, and RS-15 differ from regular repeaters in that they receive a wide range of frequencies on one band (say, between 145.860 and 145.900 MHz) and linearly retransmit those signals on another band (perhaps between 29.360 and 29.400 MHz). This means you can transmit on, for example, 145.880 and simultaneously hear your retransmitted or "downlink" signal on 29.380 MHz. A frequency-relation chart for RS-10/11's Mode A transponder is shown in Figure 5-13 and may help to make this a little clearer.

Looking at the figure, you'll notice that a "robot operator" is shown on 29.403 MHz. This is one of the most exciting and challenging aspects of RS activities. Simply explained, the robot operator will call CQ and then listen on 145.820 for callers. Assuming you send perfect CW and another station doesn't interfere, you'll hear your call on the robot operator's "receiving end" (29.403 MHz). Type your message for a QSO on the keyboard following this format:

RS-10 de [your call] ar (._ ._.).

If you contact the robot, it will reply with:

[your call] de RS10 QSO nr [xxx] OP Robot tu for QSO nr [repeat of your QSO number] 73 sk.

The robot operator will then pause and call CQ again. Be sure to copy the time of your QSO and your QSO number. You can then send a QSL for the robot operator of RS-10 to: Radiosport Federation, Box 88, Moscow, Russia, and a spiffy QSL from the robot operator will grace your mailbox a few months later.

Sounds easy, right? Sometimes it is, but the robot operator can be a sly old fox. For example, you may be listening and have your computer system preprogrammed with the exact message, all ready to make a robot contact when the satellite comes into range...but then the robot operator is silent! You tune the transponder's regular 10-meter downlink passband and hear several stations engaged in conversations, but still no robot operator calls CQ. As the satellite orbits on past your QTH and almost out of range,

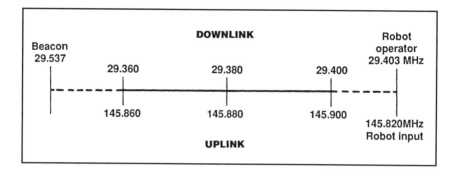

Figure 5-13. Outline of uplink/2-meter and downlink/ 10-meter frequency relations used with the Mode A transponder aboard Russian RS satellites. Note the robot operator which receives on 145.820 and transmits on 29.403 MHz. This "automated ham" is a blast to contact.

however, the robot suddenly springs to life with "CQ de RS10 robot!" Now how fast can you access your computer's program and call the robot operator? If you miss the robot on that orbit, be prepared approximately 110 minutes later for the next orbit (and better luck!). Rest assured it is fun all the way!

RS-10/11, RS-12/13, and RS-15 orbit the Earth approximately every 110 minutes, which results in approximately 13.5 orbits in 24 hours. The orbit is fixed in position, but the Earth's rotation causes each orbit to appear shifted to your west by approximately 30 degrees latitude with each pass. As a result, approximately two evening orbits and two morning orbits are usually within range of your QTH. Exact orbital information, tracking data, and daily operating schedules of the RS satellites are announced on the weekly international AMSAT Net (14.282 MHz at 1900 GMT Sundays). You don't have to transmit to copy this data—just listen to the broadcasts. (AMSAT, incidentally, stands for the Amateur Satellite Corporation). Additionally, more information on all amateur satellites, including the "RS birds," plus computer tracking programs are available from AMSAT, P. O. Box 27, Washington, DC 20044. Include a large manila with two stamps in your request for information. Computer tracking programs available through AMSAT have proven very effective and are available at reasonable costs.

Now that you have some background under your belt, let's get back to how you can use your FM handheld to transmit to the satellite on CW. First, and most important, make sure you don't have the CTCSS or "PL" tone unit in your handheld activated. It will cause your transmitted signal to produce FM rather than pure CW and will interfere with other stations. Second, assemble the following key adapter and check its operation by listening to your signal on a separate FM handheld to make sure that absolutely no audio can be heard. Again, this ensures that your rig transmits a narrow CW-only signal that will not interfere with other satellite users.

Quick-switching your handheld to transmit CW rather than FM involves disabling the rig's microphone input circuit and keying its push-to-talk (PTT) line. Let's make a plug-in adapter that will accomplish both steps at once and simply unplugs to restore normal FM operations. The parts you'll need are a small telegraph key, a three-conductor mini-plug to fit in the external microphone socket of your FM handheld, and a small assortment of $1/2$-watt resistors with values between 27 and 76 ohms. When this plug is inserted into the microphone jack, it automatically disconnects the rig's internal microphone and lets you key the PTT circuit. Most handhelds are wired so the mini-plug's tip connects to the microphone, the center ring connects to the PTT, and the back shell section connects to ground. A quick experiment will confirm if your particular handheld is wired in this way or if the ring section is microphone and the tip is PTT. Connect a rubber ducky antenna to your handheld, then use a clip lead to briefly short the mini-plug's shell to its ring or tip to determine which one causes the rig to transmit. The other connection then is the one used for the microphone (and can probably be disregarded for CW use, as we'll discuss presently). After determining which mini-plug terminal keys the PTT, insert your highest value resistor in series with that wire and the key as shown in Figure 5-14.

Typically, your highest value resistor will not allow the transceiver to transmit. Drop to a lower value resistor (for example, 56 ohms), and try to key the handheld on transmit again. If it works and the rig delivers full rated output for its particular battery

Figure 5-14. Wiring arrangement for transmitting CW/Morse code with an FM handheld. The microphone is disabled to produce a clean-carrier signal.

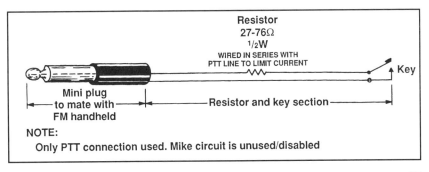

pack, solder that resistor into the line between the mini-plug and key, and your modification is almost complete. If the handheld still does not switch to transmit, continue lowering resistor values until proper keying is achieved. Our purpose in these steps is to minimize current flow through the keying circuit and to ensure that the switching transistors within the handheld do not overheat due to excess current. Some equipment accessory manufacturers may overlook this step in their "after market" external speaker mikes, but we want you to do it right the first time!

Next, borrow a friend's FM handheld and tune it to the same frequency as is set on your handheld and listen to your transmitted signal. There should be absolutely no hint of audio from your rig when you talk into the mike. If there is, short the (keying adapter's) mini-plug's shell and microphone terminal together to disable the internal mike. Double check against the borrowed handheld transceiver again to make sure the transmit audio has been muted, then get ready to enjoy some exciting CW contacts on the RS satellites. If you have an optional 5-watt battery pack for your FM handheld, now is the ideal time to put it to good use. This increase from 2 to 5 watts will make a noticeable difference in accessing the satellites.

Turnstile Antennas for Satellite Operations

Assuming you assembled the keying adapter for FM handhelds that was described above, the following antennas are all you need to complete your super-economy satellite station. These antennas are basically cross-connected dipoles set up to make your signal rotate like a corkscrew and match the roll of a satellite in space. This approach minimizes fading and produces a very effective antenna at very low cost. Furthermore, the antennas only have to be mounted a few feet above ground to produce the desired high angle of radiation (concentrating your signal upward rather than toward distant horizons), eliminating the need for beams, rotors, and tracking—yet they still work from horizon to horizon! In other words, they won't do a bit of good for local FM operating, they do a stand-up job for the RS satellites because they're in low orbits (approximately 1000 kilometers above the Earth). Fancy beam setups are only necessary when working high altitude satellites, like OSCAR 13. These crossed dipoles, or turnstiles, will work out much better for satellite use than a regular vertical or rubber ducky antenna.

The material required to assemble a turnstile for 2 meters and a second turnstile for 10 meters is approximately 50 feet of copper wire (No. 12 or 14 is a good choice), a roll of 50-ohm RG-58 coax cable (enough to connect both of these outdoor antennas to your indoor rig), insulators, solder, tape, a PL-259 connector for each cable, and a PL- 259-to-BNC adapter for mating the 2-meter turnstile to your FM handheld.

Refer to Figure 5-15 for assembly details. Notice in these top views that the antennas are mounted horizontally, with any two adjacent elements connected to (only) the coax cables' center conductor, and the other two adjacent elements connected to (only) the coax cables' shields. Element length is approximately 19 inches on each of the 2-meter wires and approximately 7.95 feet on each of the 10-meter wires. These lengths were calculated, incidentally, using the standard formula: $234 \div$ Freq. (MHz) = length of each $1/4$-wave wire/element (in feet). Since the 2-meter turnstile is smaller and easier to assemble, I suggest building it first. I'll leave selection of a center insulator to your ingenuity. Solder all wires at the coax feedpoint securely, then wrap them well with tape or Coax Seal for weather protection. Finally, install PL-259 connec-

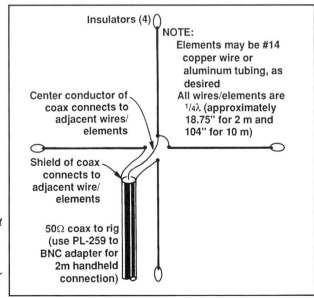

Figure 5-15. Outline of turnstile antennas for operating RS satellites. Note that two of these antennas are required: one for 2 meters and one for 10 meters.

tors on each coax cable's end (and mark which one is 2 meters and which one is 10 meters!).

Since these turnstile antennas are designed to radiate upward, they need be mounted only high enough to avoid snagging folks walking through your yard. The one stipulation I'll add is to try to mount them in the clear rather than amid foliage, which can absorb RF energy like a sponge. Think creatively: the turnstiles might even be mounted low on a roof with a short vent pipe providing center support. Finally, separating the 2-meter and 10-meter turnstiles by at least 5 feet will ensure minimum crosstalk or desensing of rigs during operation.

Ready to try your new antennas? Listen to RS-10 during an in-range orbit/pass, then try transmitting your own signal and listening for your downlink/relayed signal. RS-10 will orbit over your QTH once or twice in the morning and once or twice in the evening, trekking from sound to north in the morning and north to south in the evening. The orbits will even shift westerly a small amount each day, which means that if there's an overhead pass on day one, the orbit will be to your west on day two. As the orbital trek proceeds on to the west, the track of the east pass/orbit will move closer to your range/area. Now fire up your complete system, brush up on your CW, and go for the satellite contacts. They're guaranteed to expand your amateur radio fun tenfold!

As you'll surely agree, this chapter described a wide variety of handy and unique projects applicable to VHF rigs. I trust they'll increase your enjoyment of amateur radio and inspire you to further investigate satellite operation. There's a fascinating and multifaceted world of amateur radio waiting for you, so keep on expanding your horizons.

CHAPTER 6
Big-Time Accessories for Your HF Mobile Rig

This chapter is dedicated to our amateur radio friends who commute to work, who often find themselves traveling the open road, or anyone else who appreciates the benefits of a good mobile setup. Here I describe several projects to make hamming from your vehicle more enjoyable and to ensure that your station on wheels pumps out a great signal. I've included some ever-popular projects like a neat homebrewed antenna and tips on setting up a good ground system, plus some unique goodies like a dual antenna system and mobile CW electronic keyer to help you move into the big league and have even more fun hamming in the great outdoors.

Newcomers may envision mobiling as a strictly "in-motion" operation, but that's not necessarily the case. Indeed, a mobile setup is the ideal way to carry your ready-to-use station along on vacations or weekend visits with relatives. That's right—even a brief 15 minute operating stint from a parking lot or driveway can turn an otherwise boring outing into loads of fun. I vividly remember, for example, carrying a small QRP transceiver and quick-mount antenna in a rental car during a recent trip to a Gulf Coast beach. I snapped everything into place while I was parked in a shopping mall and made one contact, then left the rig propped against the seat for later use. Later that evening, while parked in the rental condo's lot, I switched on the rig a second time and worked several DX stations in just 20 minutes. Needless to say, that mobile approach was much easier than carrying gear into the condo, putting up an antenna, etc.

Another good reason to have a rig "at the ready" in your vehicle is so you'll always be prepared for impromptu operating. In fact, this capability seems to be the secret ingredient to DXing success. It's the classic story of band conditions and DX always being best during the least opportune time, say, when you're away from your home shack. But a good mobile setup helps you avoid that dilemma, and, as I always say, "it's better to have it on hand and not use it than to need it and not have it." So, go for it!

Homebrew a High-Performance Mobile Antenna

Perhaps you're ready to go HF mobile and need a hot antenna that will really reach out and let you work distant areas with ease, but you've been held back by the high cost of the commercially available antennas and mounting problems associated with modern bumperless autos. This project is the perfect answer! It describes a quickly assembled mobile antenna that's extra tall to deliver a whopping signal, is lightweight, and has very low wind resistance. The antenna can be assembled for operation on any band between 80 and 10 meters and the upper sections can be swapped in a flash for multiband operation. In addition, you can also "build in" a 2-meter antenna if you want. The "short base version" can be mated with most popular trunk lip mounts, while the "tall base version" can be supported by a strap installed below an auto's rear bumper area. An impedance-matching coil can also be included at the base to produce much higher efficiency than most regular mobile antennas can offer. In other words, this antenna project is flexible enough to fit all needs. Plus building it is kind of like taking a simplified mini-course in mobile antenna design—and you can use the knowledge you'll gain for many years to come!

Before delving into the actual antenna assembly, let's look at some of the fine points associated with radiating a great mobile signal. First of all, any mobile antenna other than a full $1/4$-wave whip for 10 meters necessarily uses a near-center loading coil to reduce its overall height. Basically, this "loading concept" can work in one of two ways: by using a multi-turn coil and a short upper whip (section above the coil) or by using a coil of fewer turns and a longer upper whip. What's the difference? Plenty! Since loading coils are rather lossy and the whip section above the coil is an antenna's main radiating area, using "less coil and

Figure 6-1. Homebrewed high-performance mobile antenna installed and ready for use on author K4TWJ's auto. A short mast section made from the "large end" of a CB whip is fitted with $^3/_8$-24 thread adapters and mated with a Hustler resonator for a higher-than-desired band. The resonator's top whip is then replaced with a tall (42-inch) "stinger."

more whip" (rather than vice versa) improves signal radiating efficiency significantly! Second, for best results, the antenna's loading coil should be raised (by the below-coil mast section's height) above the auto's roofline. This arrangement also improves efficiency. Finally, mobile antennas typically exhibit 3- to 14-ohm, rather than the expected 50-ohm, base impedance, and a base-matching coil further improves efficiency. Now the good news: all of these desirable qualities are integrated in the mobile antenna described below!

Five photos of our "super-duper mobile antenna" are shown in Figures 6-1 through 6-5. Most of the parts required for assembly are available at your local amateur radio dealer or hamfest fleamarkets, or you can homebrew many (or all!) of the parts to cut costs. The "short base version" for trunklid mounting uses 19 inches of a 108-inch CB whip's large diameter lower section as a mast. It's fitted with $^3/_8$-24 thread adapters on both ends and is combined with a Hustler resonator/loading coil. The resonator's standard/short top whip is replaced with a Ten-Tec (or equally thin) 42-inch-tall top stinger. The "long base version" for under bumper mounting is similar, except that a 4.5-

Figure 6-2. A close-up view of base impedance matching coil. A spade lug at the bottom of the coil fits between the antenna's lower end and mount, and a tap near the middle of the coil connects to a short jumper to the auto's frame/ground.

foot Hustler mast or a 5-foot section of $1/2$-inch aluminum tubing with screw stock wedged in each end is used. The base matching coil can also be "homebrewed" or purchased ready-made from Lakeview Company (3620-9A Whitehall Road, Anderson, SC 29624; Phone: 864-226-6990). The coil is 2.5 inches in diameter and 2.5 inches tall, with nine turns total. The coil's lower end is connected to the antenna's base where it screws into the mount and is tapped three turns from that "lower end" for matching impedance on 15, 17, 20, and 30 meters. A tap five turns from the bottom usually matches impedance for 40 meters, and a full nine turns usually matches impedance on 80 meters. The matching coil is usually not required for 12- or 10-meter operation.

A general outline for assembling this antenna is shown in Figure 6-6. A Hustler RM-15 resonator and a 42-inch (Ten-Tec type) top section is used for 20-meter operation. A Hustler RM-20 and similar 41-inch (as measured from top of resonator to tip of antenna whip) top section is used for 30-meter operation. An RM-30 resonator and an approximate 48-inch whip can also be used for 40-meter operation. In the latter case, a small chrome junction unit, approximately 2 inches long, is used to mate the Ten-Tec whip with another stinger of similar diameter to achieve the longer top section length. These junction items are usually available from hobby stores nationwide.

This short mast or trunklid-type antenna is mainly designed for big signal operation on 40, 30, 20, 17, and 15 meters. It can also be used on 12 or 10 meters in ultra-short form (4-foot total height and perfect for slipping into garages) by using a standard/stock RM-10 or RM-12 resonator and extending its supplied (short) stinger an approximate $3\text{-}1/2$ additional inches "out of the resonator (total stinger length is then $10\text{-}1/2$ inches). A better bet in the latter cases, however, involves mounting the antenna at bumper level, using a 4.5- or 5-foot mast section and replacing the standard/short stinger with a 42-inch-tall top section, as previously discussed for 30 and 40 meters (or even 80 meters). In this case, you can add a jumper around the resonator so the antenna's overall length is approximately 104 inches: a full $1/4$-wave on 10 meters and a really terrific antenna!

The previously mentioned resonator jumper can be quick-assembled as follows. Remove approximately 8 inches of shield from an old piece of coax cable and solder one end to a large lug that will slip over the screw attaching the resonator's lower end. Then sol-

Figure 6-3. Ultra short version of a homebrewed mobile antenna also works any HF band, but stands only 4 feet tall—perfect for slipping into garages. It uses a 19-inch base and coil, like that shown in Figure 6-1, but a shorter stinger.

Figure 6-4. The top section of the mobile antenna is unscrewed, leaving a hefty $1/4$-wave whip for working 2 meters in high style.

der an alligator clip on the strap's other end for clamping it above the resonator (on the chrome part). Plan ahead (say you use an RM-15), and you can achieve a two-band antenna that will operate both 10 and 20 meters by simply inserting or removing the jumper and disconnecting the base impedance matching coil's jumper.

Before discussing initial checkout and tuning of this antenna, I'd like to share with you some of the theory that was integrated in its design—this way, you can visualize exactly how the antenna works and tailor its tuning to your own needs. You can also wind your own center loading coil if desired and fabricate its center mount from fiberglass or other on-hand materials. First, I reasoned that a small loading coil (like an RM-10) could be used to make a 20- or 30-meter antenna, but the top stinger section would probably be excessively long (about 8 to 15 feet!). The most logical stinger I could find that would fit in the resonator and fit under street lights when in motion was 42 inches. So, I inserted several inches of its wide end into an RM-15 and used an MFJ-204B antenna impedance bridge and small portable digital frequency counter to check that combination's resonant frequency. It measured 20 ohms at 14.8 MHz, so I stretched the 42-inch stinger all the way to within 1 inch of its tip, and its resonance moved to 14.250. I then added the base matching coil and moved its tap until impedance measured at the auto transceiver was 50 ohms.

Substituting other resonators and top stingers produced resonance at a variety of out-of-band frequencies, so I again added "more stinger" to lower the frequency of resonance or a shorter stinger to raise the frequency of resonance. Read that last sentence again. It means you can assemble this antenna to operate on any frequency desired! If you don't own an antenna impedance bridge, however, simply follow my previously discussed dimensions and use an SWR bridge to determine if you should lengthen the stinger to lower the antenna's frequency of lowest SWR or shorten the

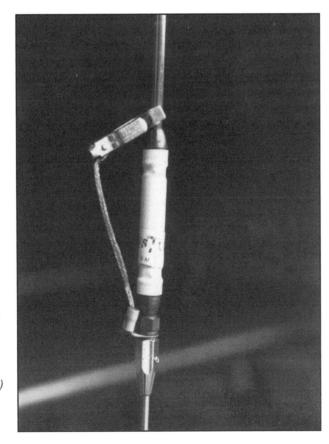

Figure 6-5. Homebrewed jumper installed between the bottom of the resonator and the top area of the whip. This arrangement permits using a full-size (104-inch) whip as a full $1/4$-wave antenna on 10 meters.

stinger to raise the antenna's frequency of lowest SWR. Likewise, you can wind a homebrew coil of "X" number of turns and mate it with an on-hand stinger, then check resonant frequency with an MFJ antenna bridge and adjust the coil and/or whip length until resonance falls within your preferred band and range. This procedure is actually even easier to achieve than it sounds. In reality, setting a homebrewed antenna of unknown resonant frequency to a desired band and frequency takes less than five minutes.

Assuming you make the "short mast version" of this antenna, the lower section is approximately 19 inches tall. You can thus remove the upper resonator-and-whip section and bingo: you have a 2-meter $1/4$-wave antenna! If you assembled the taller base section version, consider painting the aluminum tubing a color to match your car. Whatever direction you choose, rest assured that this antenna pumps out one terrific signal. I use one almost daily, and everyone says my mobile signal sounds like a base unit! You, too, can enjoy such high performance!

Radio-Active Antennas

This simple project is dedicated to the fun-loving types among us who are concerned that life is becoming a little too serious. It describes the classic quick-trick of adding a small neon lamp to a mobile or home antenna so it winks and blinks against the evening sky according to your transmitted signal. This effect of a "glowing antenna" doesn't degrade your rig's performance, and it really captures attention. Furthermore, you can look right out your window and watch your signals departing for distant lands! Ready for some genuine old-time amateur radio fun and excitement? Read on!

Briefly explained, the blinking antenna effect is achieved by clipping a small NE-2 or similar neon pilot lamp to the tip of your mobile whip or home station's antenna (see Figure 6-7). If desired, small NE-2 neon lamps can also be connected to wire ends of home antennas, such as dipoles, inverted vees, or even the driven element of beams. And you can experiment with adding an NE-2 to your 2-meter antenna, but it may not light unless you're running more than 50 watts of output power. The RF energy from your transmitted signal ionizes the lamp's neon gas and causes it to flicker with SSB voice peaks or to blink with Morse code. In fact, amateur friends viewing your antenna could visually copy your Morse according to flashes while transmitting. Impressive? You bet!

Only one of the neon's wires needs to contact your antenna for ionization at power levels of 60 to 100 watts. Just clip the antenna in place, stand back, and watch the effect! During holiday seasons, you might

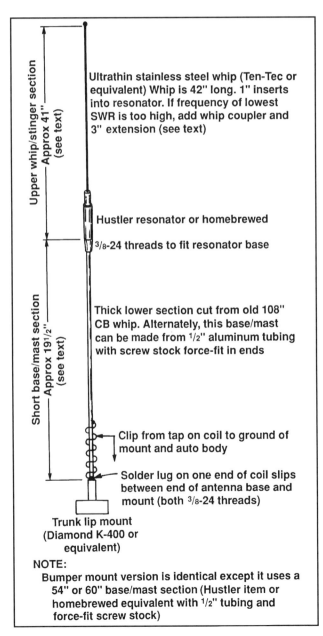

Figure 6-6. Assembly outline of the high-performance HF mobile antenna. This big-signal radiator can be made in several ways and tailored for operating any desired band.

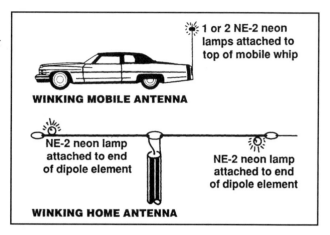

Figure 6-7. Two examples of antennas that wink and blink when you transmit SSB or CW.

also want to spray paint the lamp with translucent colors to create a festive appearance. Or you might consider trying some of the popular plug-in night lights as NE-2 substitutes. Check these items carefully before purchasing, however, to make sure you're getting a genuine neon lamp rather than an "incandescent imitation." Use your imagination, and enjoy the neat effects of a soft winking antenna as you DX the bands!

Phased Antennas: A Serious Mobileer's Delight

OK, outdoor hamming enthusiasts, this one-day project makes for terrific fun on the open road. It's a dual antenna system that effectively doubles your transmitted power strength while boosting reception accordingly, and it works with all types of HF mobile antennas. You can even use it with two identical homebrewed antennas like those discussed earlier in this chapter. You can set up the dual antennas to work 10, 12, 15, 17, or 20 meters, or even install antennas for separate bands in each rear mount and use an interior coax switch for instant band changing. Furthermore, you can plug the coax feedlines into two separate HF transceivers installed in your vehicle and operate two bands simultaneously. If that doesn't get your adrenaline pumping, check your pulse!

An assembly outline of the phased antenna system is shown in Figure 6-8. A pair of Hustlers, Ham Sticks, Outbackers, or homebrewed antennas are mounted on each side of your auto's rear bumper or quarter panel, and connected to two pieces of coax cable. These two cables should be identical in length (17 feet exactly) and type, and preferably cut from the same master spool of cable, and without any splices whatsoever. I'd strongly suggest low loss RG-8X, with the ideal choice of cable being RG-8X Marine Grade. You could use also regular RG-58U cable, but just be careful to route it so the center conductor doesn't become crimped by seat edges, etc. If your vehicle is large and 17 feet isn't enough length, you can use a 21-foot length without any adverse effects. Remember, however, both lengths of cable must be *exactly* the same.

Connect each cable to its antenna socket or base matching network in the same manner. Make everything an exact duplicate. If one antenna has a home-brewed matching network on its base, build an identical matching network for the other antenna. Likewise, all ground connections should be mirror images. Route the two cables through similar paths on each side of the car to the front/transceiver area. Use an 83-1T "Tee" connector at the back of your HF transceiver to hook up the phased antenna system; then, when you prefer a non-phased system, simply unplug one antenna's cable and remove the whip from the car. Alternately, antennas for separate bands can be installed in each mount and their cables switched or swapped at the transceiver's back for dualband operation. The only restriction is that you must avoid having antennas for the same band in both rear mounts and using only one of the antennas. The "unused" antenna will probably influence the SWR of the "used" antenna.

Initial checkout and tune-up of the phased/dual antenna system involves connecting only one of the antennas to the transceiver (through the 83-1T) and adjusting its stinger and base-matching section for lowest SWR. Now unplug that antenna and plug in the other antenna (only). Adjust its stinger and base-matching unit for the same value of SWR as the first antenna. Next, plug both antennas into the 83-1T "Tee" connector (which will be plugged into your transceiver, naturally!). The SWR may increase slightly, but it should not increase significantly (above 1.7:1). Now switch on the automatic antenna tuner in your transceiver, and use it to match the dual antennas and produce a perfect 1:1 SWR. If your transceiver doesn't have a built-in antenna tuner, a small and inexpensive manual tuner (like an MFJ-945D unit) can be used. If the antenna system's SWR is less than 1.5:1, however, you can probably forego the tuner altogether and enjoy using the antenna system directly.

These phased antennas deliver approximately 3 dB gain in-line with your vehicle's front and rear, with a slight null noticeable off to the sides. Remember that directivity characteristic when chasing DX and enjoy the benefits of using one of the hottest mobile antenna systems on wheels!

Installing a Great Mobile Ground System

Have you ever wondered why some HF mobile setups seem to pump out a great signal and experience minuscule ignition noise problems, while others using similar model 100-watt transceivers are fortunate to make even an easy contact? The difference may be due to band propagation ("skip" favoring the mobile station), or it could be that a clever operator has added a super-effective ground strap system to the vehicle. How much improvement could adding ground straps make in your own mobile setup? Let's put that in perspective with an easily visualized example. Remember this chapter's first project explained how reducing the size (inductance) of an antenna's loading coil and using a longer top whip could significantly improve antenna performance? Well, additional improvement can be achieved through effective grounding, which will further increase the overall efficiency of your mobile antenna! That's a two-for-one gain definitely worth pursuing! Ready to become a genuine big league mobileer? Terrific! Let's get started!

The main items you need for this project are a roll of copper grounding foil (available through larger amateur radio suppliers like Ham Radio Outlet) or some flexible grounding braid (like the shield removed from an unused length of large/RG-8 coax). You'll also need an assortment of spade lugs, alligator clips, some sandpaper, and your trusty voltohmmeter to guide and check your work.

This project is a combination of "learning and doing" that you can finish right away or work on one step at a time. Let's begin by describing some often-overlooked automotive pitfalls you can correct through grounding. This grounding process will accomplish two feats at once: improve your radiated signal and reduce ignition noise. But I should also point out that some of the following steps are applicable to all mobile installations; all are applicable to some installations; but all are not necessarily applicable to all installations.

Let's tackle the most rewarding improvement first, which is ensuring that your transceiver and antenna are securely connected to your auto's frame for a solid ground system. First, install a fresh new ground strap between your antenna mount's ground lug (where the coax feedline's shield connects) and the auto's frame or main body (depending on whether you're using bumper or trunk lip mounting). Measure the shortest length of strap necessary to connect the two points, then install a spade lug on one end for the antenna mount and a larger lug for mating with a convenient body bolt on the other end. You may need to drill a small hole in part of the auto's frame at this point, but look closely before setting drill to metal, however, as a body bolt or similar attachment point can usually be spotted by careful investigation.

Use fine sandpaper to remove caked road grime, paint, undercoating, etc. where your lugs make contact with the mount and vehicle body. Use star washers beneath the lugs to "bite into metal surfaces," and ensure good electrical contact. This step is particularly important for bumper-level mounted antennas because they're susceptible to rapid deterioration from the elements. In the case of trunk lip mounts, attaching an extra ground strap under one of the mount's set

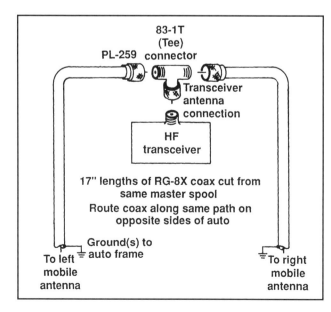

Figure 6-8. Outline of phased mobile antenna system. This setup delivers a band-commanding signal and also improves reception of weaker stations.

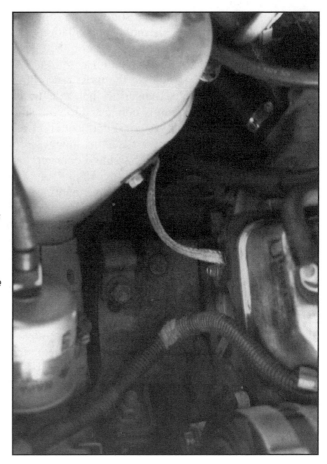

Figure 6-9. An example of how flattened coax shield or flexible braid is installed and used to minimize auto ignition noise while improving grounding efficiency. In this case, spade lugs on each of the strap ends were slipped under bolts on the auto's engine and body. Several more ground straps are included in this installation, but we could not snake a camera and flash unit under the auto's body to photograph them!

screws and then scraping some paint from around a trunk lid mounting bolt should be adequate. Most vehicle manufacturers paint the trunk inside, paint its hinges, then install mounting bolts. If the paint was dry when the trunk lid was attached (typical), the trunk lid can be electrically insulated from the rest of the body (your antenna will work with a slightly high SWR, but the overall system lacks a proper ground). Likewise, antennas mounted on bumpers that are insulated from the auto by shock absorbers lack proper grounding. Rest assured your extra ground straps will make a significant improvement regardless of the manufacturer's design.

Next, install an independent ground strap between your vehicle's body and the transceiver's cabinet. Yes, I know your rig is connected to the battery's negative terminal and that constitutes a "ground," but that's an electrical DC connection, not a proper RF path. Again (and most important) add an extra grounding strap from your vehicle's body to your transceiver's cabinet. This is easily accomplished by measuring another length of flexible braid and attaching an alligator clip to both ends. Look under the seat and you'll see one or more bolts holding the seat to the frame. Connect your alligator clamp to one of those shiny bolts; the strap's other end can then be quick-clamped to your transceiver's ground lug for easy rig installation and removal.

Now let's check your work for accuracy. Step to the rear of the vehicle and short the antenna's whip (which is connected to the coax cable's center conductor) to its mount's ground lug with a clip lead. Now move to the interior and use your ohmmeter to check overall antenna system conductivity as follows. Place one ohmmeter lead on the antenna coax cable's PL-259 center conductor (be sure you touch only the center conductor and not the shield). Touch the other ohmmeter lead to the transceiver's grounding cable emerging from under the vehicle's seat. Your transceiver should not be connected (or even *in* the vehicle, for that matter) for these checks. Using this method, the coax cable, antenna mount, ground lugs, and ground straps are all checked for continuity simultaneously. Your ohmmeter should read less than .5 ohms resistance; if it reads higher, rework your ground connections one at a time using the ohmmeter to locate poorly connected ground straps. After achieving an overall resistance of less than .5 ohms, you can remove the antenna's rear jumper and recheck for an open circuit condition. Assuming everything checks out properly, use Coax Seal or other silicone-type protector to weatherproof all of your ground strap connections.

You now have a high-performance ground system that will significantly improve on-the-air operation. But you're not done yet—you can continue making improvements to "soup up" your setup.

Your engine is the primary noise generator and is usually supported with hard rubber motor mounts. The engine's tailpipe is also usually supported or hung below the vehicle by thick rubber brackets and acts as an underbody antenna for the noise generated by the motor, terminating near your antenna's mount. Adding ground straps between the engine and vehicle's frame and between the tailpipe and frame will noticeably reduce ignition noise (see Figure 6-9). I encourage you to ground the engine on both sides and on its front and rear. Likewise, try to ground the tailpipe in at least three places (one of which is near its end and your antenna mount). If you experience problems with RF energy affecting the computer system, here are a few tricks proven to minimize that interference.

First, route your transceiver's DC cable through the vehicle's left side and directly to the battery. Likewise, route your antenna's coax on the left side to the rear mounting area. Add aluminum foil as a shield around the vehicle's microprocessor which is usually mounted in the right front kick panel. Insert a sheet of plastic between the aluminum foil and the microprocessor, so you don't short any circuitry within the processor, then add a ground strap between the aluminum foil shield and the vehicle's body. Finally, adding snap-on toroids to all rig lines (DC cable, external speaker, etc.) will minimize RF feedback and produce a super-smooth operating system.

I hope you've found this chapter's projects helpful in setting up and improving your existing mobile station. Hamming on the open road is a lot of fun, and every minute you invest in perfecting your setup will be returned with improved results!

CHAPTER 7
A Potpourri of Fun Projects

Now here's the really fun chapter of this book. It's a super collection of electronic projects guaranteed to hook you on homebrewing. Scan through the following pages and you'll surely agree that there's something here for everyone: new amateurs, old pros, and even general electronic experimenters! There's a blinking eyes troll doll to amuse your friends, a pocket transmitter for ham club and hamfest fun, an old-time telegraph system to help you recreate the old wild west days of Morse, and much more.

Each project is an example of homebrewing at its best and most fun. They're also intended to inspire you to try out your own ideas, so feel free to be creative in layout and packaging. Avoid modifying the actual circuit designs, however, until you've confirmed proper operation of a project as described. This way, you can always return to square one if a certain change didn't work. That's enough "electronic philosophy" for now. I'm sure you're anxious to get started assembling!

Blinking Eyes Troll

Here's a project you can finish in one evening and that makes a terrific conversation piece for your ham shack as well as a unique gift. Simply explained, you'll be using an NE-555 multivibrator IC to alternately flash two LEDs inserted in the eyes of any doll or stuffed animal of your choice. Of course, they must be seen in a dimly lit room to be fully appreciated!

Two of the critters that I homebrewed are shown in Figure 7-1. The item on the left is exceptionally small,

Figure 7-1. Two blinking eyes toys assembled using LEDs and the NE-555 circuit described in this project. The troll doll on the left is 2 inches high (less hair!) and has a hollow plastic body. On the right is a paper maché owl with hollow inner section. Both of these critters are real attention grabbers!

so it proved quite challenging to construct. Rather than using a perfboard, I assemble the circuit "open air" style with a capacitor in one of the troll's arms, resistors in the middle, and the IC in the little guy's tummy. After assembling the circuit and tape-wrapping interconnecting wires, I removed the troll's hair to access the doll's inside for installing the circuit. I then added a battery clamp to hold a 9-volt battery that sits behind the troll (and supports him upright), and glued its hair back in place. The item on the right of Figure 7-1 was originally an incense burner (holder), but its inside was open so sliding a pc board plus battery in the bottom was a cinch. The eyes of both dolls were drilled out and small LEDs inserted.

All of the components required to assemble a blinking eyes troll (including LEDs and perfboard) are available from RadioShack. If there's a hobby/craft or discount store in the same shopping center, buy the doll at the same time. This way you can visually compare the electronics and the troll's interior space to make sure your assembly will fit. Be sure to allow room for inclusion of a standard size 9-volt battery, and you should be right on track. Considering the wide variety of such toys, the only problem you'll face should be deciding which one to buy!

The NE-555 circuit is shown in Figures 7-2 and 7-3. You should have little difficulty assembling it on a 1.5-inch-square piece of perfboard. Even if you must make the electronics smaller, perhaps to fit in a miniature doll, I'd still suggest first building a larger prototype to gain experience (just use it for an in-shack companion). Then you can assemble the smaller version using $1/4$- or $1/8$-watt resistors instead of standard $1/2$-watt resistors. Lay out all the components on the corner of your perfboard (it should be larger than necessary to fit everything), mark where each component will be mounted, then trim or cut the perfboard to size and assemble/wire the finished product. Perfboard can be cut with a small coping saw, but I prefer to simply use diagonal cutters and trim to size. Align the diagonal cutters with the holes in the board, then "snip"! The only disadvantage of using cutters is that you'll occasionally snip off larger or smaller pieces than desired and will have to start over. But the advantage is that diagonal cutting is quick!

Use my outline (Figure 7-3) as a quick and easy assembly guide. Notice that this is a top view of the IC, and its indentation or referencing dot on the left side indicates pin 1 of the IC. Notice, also, that the capacitor's negative end connects to pin 1 and the battery's negative terminal. I leave the choice of including an on/off switch or simply connecting and disconnecting the battery for operation to your preference.

Essentially, the NE-555 circuit is a self-oscillating multivibrator with its frequency or "blink rate" deter-

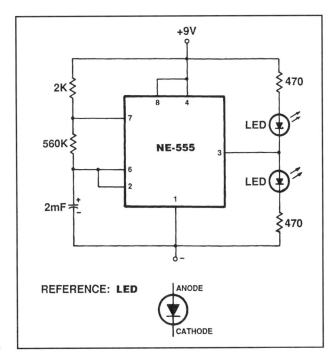

Figure 7-2.
Schematic diagram for the electronic circuit used in the blinking eyes troll.

Figure 7-3. Pictorial diagram and parts layout guide for the electronic circuit in the blinking eyes troll.

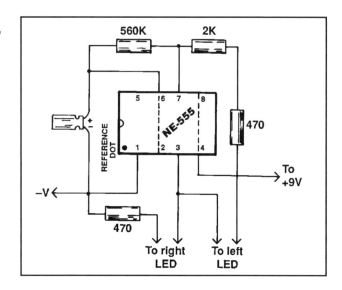

mined by the value of the 560-k ohm resistor and the 2-mfd capacitor. Reducing the size of the capacitor or resistor will increase the blink rate, and increasing the size/value of either will reduce blink rate. The 470-ohm resistors serve as loads for each of the LEDs and maintain current flow at a safe level.

The common method of marking the cathode in LEDs is shown in Figure 7-2, but, if you're unsure of the markings, I would suggest the following circuit checkout procedure before installing the LEDs. First, apply 9 volts of power to the circuit and connect your ohmmeter (with 5 volts DC scale selected) between the NE-555's pin 3/output and pin 1/negative terminal of the battery. Assuming proper operation, the meter's needle will flicker approximately four times per second. If the meter's needle doesn't move, switch to the milliampere scale (100-mA range should be fine), and connect the meter's leads between the battery's negative terminal (negative meter lead) and the "master connection point" at the junction of pin 1, the 2-mfd capacitor's negative lead, and the 470-ohm resistor (positive meter lead). Approximately 10 mA of current should be indicated. If no current is being drawn, recheck your wiring; you'll probably find an open connection in the circuit. If excessive current is being drawn, you've probably shorted two wires together accidentally.

After correcting any possible open or short circuits, you're ready to add extension wires from the LEDs mounted in the doll's eyes. A bit of experimenting is required here because there are four wires and a reverse-connected LED will not light or blink (hence a 50/50 chance). Use small alligator-type clip leads to connect leads (and reverse, if necessary) to one LED at a time, then note which wire connects where and replace the clip leads with permanent extension wires (see Figure 7-4). Add a battery clamp and the project is completed.

I'll conclude with just a few more tips to make this project extra special. Use a slow speed drill and sharp-tipped bit to avoid cracks when working with plastic or paper maché figurines. Green LEDs for the eyes give the most dramatic effect, but bi-polar green/red LEDs also produce a unique appearance. After assembling two or three of these little gems, consider making one with three or four NE-555 circuits and six or eight blinking LEDs installed on something appropriate, say a toy turtle's shell or even interconnected (with ultra-thin No. 24 wire) to make an unusual YL hair decoration. Your only limit is your imagination. Let your creativity flourish!

Electric Wiener Roaster

Many years ago, I quick-assembled a simple electric hot dog cooker using little more than a handful of nails, a support board, and a hank of hookup wire. The project was used for summer cookouts, was admired by visitors, and then forgotten...until an almost identical commercially made version surfaced a few years later. So now, for the first time, I'm proud to share with you *my* assembly details of this hair raising mini-

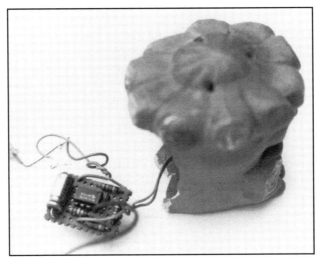

Figure 7-4. Quick-connecting LEDs to NE-555 circuit (and reversing wires as necessary to achieve blinking) is easier before LEDs are glued in the toy.

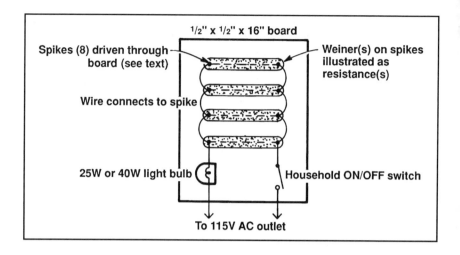

Figure 7-5. Layout of the electric wiener roaster showing wiring.

project. Why do I say "hair raising?" Mainly to emphasize that this roaster works by applying 115 volts AC to wieners inserted between exposed metal pegs and that it should never be handled when "cooking" (plugged into an AC outlet). Be sure it's unplugged, then slip three or four wieners into place, insert the AC plug into a wall outlet or extension cord for five or six minutes, and STAND CLEAR, then unplug and move the wieners to awaiting buns. Sound interesting? You bet!

A general assembly outline of the electronic hot dog cooker is shown in Figure 7-5. My original version used 3-inch-long nails rubbed to a shiny finish with sandpaper and driven through a $1/2$ x 12 x 16-inch board. This setup is definitely not FDA approved: a more modern and socially acceptable version would use stainless steel spikes, silver tips removed from home cooking utensils, or similar (here's another opportunity to use your own ingenuity in the design).

Interconnecting wires are routed between the spikes that will hold the wieners, as shown in Figure 7-5. Wrap one wire twice around each (left side) spike, then twist it tight (like tie-wrapping a trash sack); continue on to the next spike and repeat the process. Follow a similar procedure for the right side's spikes and wires, then connect a surface-mount light socket in series with the left side's wire and a switch in series with the right side's wire. Screw a household 25- or 40-watt lightbulb into the socket, add a 6- or 12-foot AC line cord to the setup, and it's ready for testing.

The device cooks the wieners through the heating effect of their internal resistance. Cooking thus begins slowly (because the wieners are dry and cool), then increases exponentially as they warm. The lightbulb serves as a "cooking monitor" and safety device; that is, the bright, uncovered light is a conspicuous "hands off" indicator, plus it also sidesteps popping a home circuit breaker if the metal spikes are accidentally shorted.

If you want, you can also make a "deluxe version" of this project by adding a sturdy plastic top cover (maybe with a switch interlock) to minimize shock potential to unfamiliar users. In other words, handle the cooker with care and respect...and enjoy the dogs!

Soil Dampness Tester for Home Plants

If you enjoy the cheerful atmosphere produced by home plants, but have difficulty determining when they need watering (especially members of the cactus family), this project will help you avoid that familiar "drooped over" look. It's a neat test probe that you can slip into the plant's pot to determine soil dampness right at root level. If the tester's LED lights green, the plant is adequately watered; if it doesn't illuminate, the little critter needs a healthy drink. Sound convenient? Indeed it is, and you can build it in less than an hour.

The circuit diagram and an assembly outline are shown in Figures 7-6 and 7-7. A 9-volt battery, 220-ohm $1/2$-watt or $1/4$-watt resistor and an LED of your choice will be wired in series and connected to the inner and outer conductors of a thin 12-inch-long coaxial probe. The probe's outer conductor is exposed along its full length and its tip/center conductor extends approximately a $1/2$ inch past the far end's tip. The tester's circuit thus remains open (which means the LED doesn't light and no current is drawn from the battery) until a low resistance connection is measured by the probe. That low resistance connection occurs when the probe is immersed in damp soil; dry soil will not complete the circuit or light the LED. Since current is not drawn during non-use or "standby," an on/off switch isn't needed and battery life is exceptionally long. In my particular unit, for example, I used a 9-volt alkaline battery, and it's still working fine after at least seven years.

A picture of my own quick-assembled soil tester is shown in Figure 7-8. I used a small paper board matchbox wrapped with several layers of electrical tape to house the battery, LED, and resistor. I just

Figure 7-6. Circuit diagram of the plant soil tester. Simplicity is the key.

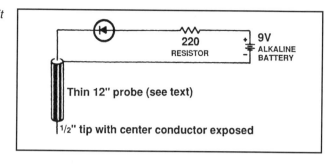

snapped a clamp on the top of the battery and wired the components in using their leads. I forced the LED through a small rubber grommet and inserted it through a hole punched in the matchbox's end, then forced the probe into the box's other end. The mechanical stability wasn't too impressive, so I just wrapped the whole thing with electrical tape—after all, I didn't plan on changing batteries for a long time!

The most challenging or time-consuming part of this project is, obviously, finding a suitably thin probe. At a household variety store, I found a small piece of temperature-resistant coaxial cable used in electric ovens. Any similarly stiff, small diameter concentric-conductor cable can be substituted here. Or consider making a probe from small RG-174 cable as follows. Remove the cable's outer jacket, then use a small (25-watt) soldering iron to carefully tin the outer conductor with a very thin layer of solder. Cut the cable's tip at one end and connect the tester's wires. Cut the cable's shield on the other end and remove insulation from the center conductor (approximately a $1/2$ inch), then tin that conductor so it's stiff. If necessary, practice on a couple of pieces of cable

Figure 7-7. Assembly outline of the plant soil tester. Layout of the parts is not critical.

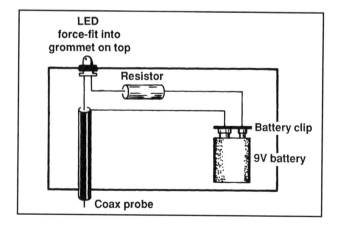

before making the finished product. Be patient and work carefully, and both the circuit and probe will go together quite easily.

Using the plant tester is a cinch. Just insert the probe into a plant's soil and notice the LED. Since soil is usually damper at the bottom of a pot than it is at the top, the little tester works great. I could have gone a few steps further and added digital readout, etc., but I considered that overkill and stuck with simplicity. But maybe you have some ideas for clever expansions, eh?

The Micronauts: Two Just-for-Fun Pocket Transmitters

This double feature project is dedicated to our ham friends who like to add a bit of fun to their regular amateur radio activities. It describes a couple of mini-rigs you can quick-assemble from readily available parts, carry in a shirt pocket and use to really surprise "low band" enthusiasts at club meetings, hamfest equipment displays, etc.

Picture, for example, the following scenario. You walk up to an operator behind a receiver or transceiver and ask if he or she hears a station calling on 3.58 or 7.160 MHz (while noting the operator's own callsign). You then begin sending Morse with your hidden key while the unsuspecting operator tunes in your signal ("What!...how are you doing that...what's going on here!"). The possibilities of this impromptu DX gag are limited only by your imagination, but there's more. Either one of the mini-rigs can be expanded into a full 300-milliwatt transmitter and combined with a BFO-equipped shortwave receiver for portable QRP operation. Then the excitement really skyrockets. Ready for your first exposure to the fascinating world of milliwatting? Read on!

The first item is a one-transistor, crystal-controlled transmitter that you can assemble on a piece of perfboard or build open-air style using small components stuffed into an oversized pen case. I'm going to describe the latter approach, but I'll use general terms so you can apply the concept to either layout or even adapt it to your unique packaging.

A photograph of my mini-rigs is shown in Figure 7-9, and their circuit diagram is shown in Figure 7-10. The pen houses a one-transistor transmitter with built-in 12-volt battery, miniature HC-18U crystal, and top-

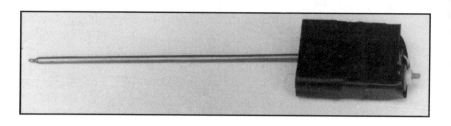

Figure 7-8. This quick-assembled soil tester is built in a paper board matchbox wrapped with black electrical tape. Top-mounted LED was force-fit into a rubber grommet, then a white Teflon washer was glued to the LED and grommet for a "finishing touch."

mounted pushbutton/key. The antenna connection is a simple wire extending from the pen's bottom, which allows me to connect it to a 3- or 4-foot flexible wire for portable operation or to an outdoor antenna for big-time QRP milliwatting.

The transmitter's circuit uses a popular 2N2222 or 2N3904 transistor in a basic oscillator arrangement. Power is supplied by a tiny 12-volt cigarette lighter battery, which can be obtained from a jewelry store or RadioShack. This particular battery is slightly smaller than a regular "N" 1.5-volt cell. You simply remove its outer wrapper, slip it into the cap of the pen, connect the top pushbutton/key to the circuitry with very small wire slipped beside the battery, and wedge everything tightly within the pen's cap. You then assemble the transmitter's circuitry, wrapping all exposed wires with tape, and insert the assembly in the key's bottom area. Use an Exacto knife to whittle off part of the pen's inside area (right where the top section threads on), and you can wedge a small HC-18U crystal in this junction. Use a plastic transistor to avoid shorts. Be careful when reassembling the pen as wires from the top battery and key can twist and break or wrap around the crystal and break its wires. Small super flexible wire cut from a low-cost earphone set works great in this application. Or you may prefer to simply glue the pen's cap in place. This will call for a bit of dismantling to replace the battery, but battery life is relatively long if the transmitter isn't used continuously. Use $1/4$-watt resistors and small diameter capacitors whenever possible to minimize sizes involved.

You'll probably notice that there isn't enough room within the pen for the transmitter's output filter section (the section shown to the right of the dotted lines in Figure 7-10). You can delete this filter, plus the emitter's bypass capacitor and even the ferrite bead/RF choke, if you don't plan on actually operating the mini-rig with an outdoor antenna. If you do want that operating option, however, you can squeeze all of the transmitter's main components (left side of schematic) into the pen and simply connect its antenna wire to the external filter section for that big-time milliwatting capability. The toroid used in the output filter section is a T-50-2 on which you wind a coil of No. 26 enameled wire, looping it through the T-50-2 core and spreading its turns evenly around the core. Be sure to scrape off outer enamel insulation to leave the shiny copper inner conductor accessible for soldering. Wind 21 turns for 80 meters, 14 turns for 40 meters, 13 turns for 30 meters, or 12 turns for 20 meters. Mate that coil with capacitors C1 and C3 for your chosen band as follows: for 80 meters use 750 pfd; for 40 meters use 470 pfd; for 30 meters use 330 pfd; and for 20 meters use 270 pfd. C1 should be 0.01 mfd for 80 meters, 470 pfd for 40 meters, 330 pfd for 30 meters, or 270 pfd for 20 meters.

The RF choke used in this little transmitter, incidentally, consists of four to six turns of ultra small No. 30 or 32 wire wound on a tiny ferrite bead (just use the smallest wire and bead you can find and it should work fine). Two good sources of crystals in HC-18U holders are Jan Crystals (you can order by telephone at 1-800-JAN-XTAL) and RadioShack (they only offer 3.58 MHz, but the price is right). Be sure to select an operating frequency authorized by your license.

No checkout or tune-up procedure is necessary with this transmitter (just press the pushbutton/key to send code!), so if it doesn't work it's probably due to a defective transistor or crystal or miswiring. You can avoid this by first assembling the rig with its full component leads, confirm its proper operation, then cut the leads to the proper lengths and reassemble the finished version (be sure to wrap any exposed wires with tape to avoid short circuits). Knowing that the project works before squeezing it into a tiny case can save a lot of frustration!

The second item of this project is a dual IC transmitter that's exceptionally small and inexpensive—you can probably purchase all the parts for less than $5 and fit it into, say, a reworked souvenir case to make...a ham radio keyfob! Please understand that I'm only making packaging suggestions, not outlining a specific design (although the rectangular dice containers found in many discount stores is a personal favorite!). Just keep your eyes and mind open when visiting gift shops and let inspiration take you! Now let's focus on this critter's circuit diagram, shown in Figure 7-11, and its general parts layout, shown in Figure 7-12.

Figure 7-9. Bearing a close resemblance to James Bond-type electronic gizmos, this miniature amateur radio transmitter is built into a pen case. The unit is fully self contained with built-in 12-volt battery, top-mounted pushbutton "key," and circuitry located in main area of pen. Might this be the world's smallest ham rig?

As you can see, our IC transmitter is simplicity itself. Indeed, only a couple of resistors and a crystal make up the oscillator section. The most common and readily available ICs in the world are used in this transmitter: the SN-7400 quad AND gate and a slightly higher current version, the SN-7403 quad AND gate. Once again, this circuit uses a small HC-18 crystal of whatever frequency you prefer and an RF choke made by winding four to six turns of fine wire through a tiny bead. The ICs are designed to operate from a 5-volt source, but you can also use three 1.5-volt alkaline cells and get good results. Or you can glue a thin metal strap on the top of the ICs to make a heat sink and increase voltage to 6 volts. Don't exceed 6 volts, however, or the ICs will immediately burn up! I should also point out that since these ICs are mass produced, they are susceptible to failure, even when brand new. One answer to this problem is to use sockets for the ICs so you can swap them until you find two good ones. Or you could preassemble the circuit on an "experimenter's block," such as those sold by RadioShack; then you simply push component wires and IC pins into the block to quick-assemble your circuit, make sure it works, then build the final version using your confirmed-good parts.

Depending on the size and enclosure you've chosen for the transmitter, you may elect to include its output filter section internally or externally. Either way, the same T-50- 2 toroidal core, the same number of turns, and the same capacitor values that were used in the pen transmitter can be used here. Clever planning can produce a very small transmitter. As an example, you might use three tiny watch batteries stacked vertically between the "L-angle" of a matchbook cover. With the matchbook cover serving as a circuit board with the components mounted on it, you'd have a "fold up transmitter" less than an inch square! Again, your ingenuity is the key!

Both of these miniature transmitters are terrific fun and you'll love demonstrating them again and again, but remember...they're also capable of *almost worldwide range* when connected to a good outdoor antenna (and assuming favorable band conditions). Indeed, I've contacted Australia several times using the little pen transmitter—and its output power was less than 300 mW! Low power communication, or "milliwatting," is thriving on HF. If you're ready for a big-time challenge, this "world down under" (the usual level of signals, that is!) awaits you. Go enjoy the fascinating world of milliwatt DXing!

Historical Telegraph System Really Works

This project is very special and, in my opinion, is alone worth the price of this book. When you're done, you'll have refurbished an old-time Morse sounder or pony line relay and interconnected it with a key and battery to make a working model of a "land line" tele-

Figure 7-10. Circuit diagram of the pen transmitter. Although this mini-rig runs only 250 milliwatts, it has a track record of contacting stations over 7,000 miles away when connected to an outdoor dipole antenna.

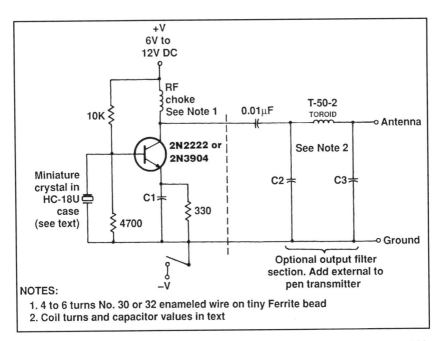

NOTES:
1. 4 to 6 turns No. 30 or 32 enameled wire on tiny Ferrite bead
2. Coil turns and capacitor values in text

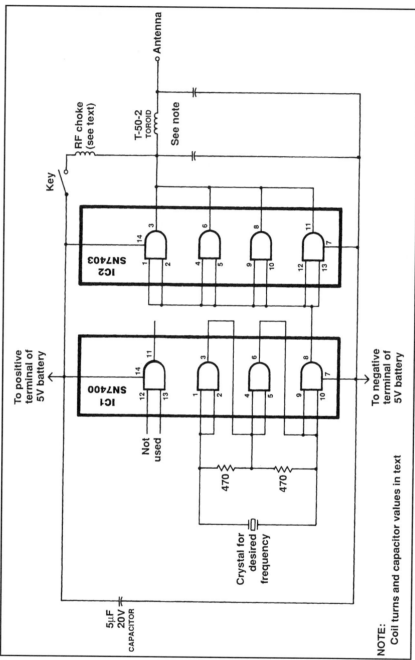

Figure 7-11. Circuit diagram of the dual IC transmitter.

graph system like those used in the old west. Setups like this were also the backbone of Western Union, the heartbeat of many railroad communications, and our country's foremost means of commercial message transmission from about the mid 1800s to mid 1900s.

Indeed, the good-old "kalick-kalunk" of continental Morse code continued emanating from classic sounders in several small town train depots around the country until the 1970s.

Sounders have essentially disappeared from communication facilities today, but they continue surfacing in amateur radio magazines want ads and hamfest fleamarkets. Used sounders, which usually cost little, are destined to become collectors' prides and increase in value with each passing year. Every telegraph sounder is also bit of real communications history that you can hold right in your hand (see Figure 7-13). Need any more reasons to seek out one (or more!) of these marvelous items? How about knowing that you'll be playing an important role in preserving our proud heritage by rescuing a classic piece of equipment from extinction? And how about being able to use it to reconstruct your own working model of a real telegraph set from eras past? An addition like that would certainly add real excitement to your home station and make quite an impression on visitors. Wow!

As previously mentioned, hamfest fleamarkets are good hunting grounds for telegraph sounders. Get there early and check out new arrivals just as they set up their tables for best results. Also, look under tables and ask questions so you don't miss exactly what you're looking for. Running your own ad for sounders and other telegraph apparatus in ham magazines may also prove fruitful. In fact, you might even acquire a deluxe sounder and key combo on a mahogany base—a terrific item after clean-up!

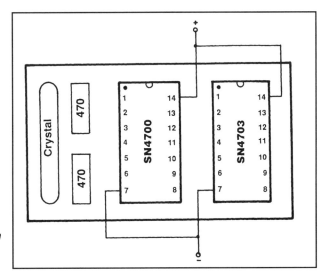

Figure 7-12. General parts layout guide for the dual IC transmitter. Sketch illustrates pinouts of ICs.

Figure 7-13. A classic telegraph sounder acquired at a hamfest (before cleanup). The unit is approximately 3 inches tall by 2 inches wide.

Study the sounder's design and you'll discover that it basically consists of two electromagnets plus a pivoting bar with adjustments for arm movement and tension. Dedicated sounders like the one shown in Figure 7-13 are usually more "upright" items, whereas the electromagnetic coils in pony relays are usually laid horizontally. A single sounder also usually has only four connections (two for each coil) whereas a pony relay usually has six connections (the other two for keying an additional circuit). Trace the electromagnets' wires to the base-mounted connection screws then use your ohmmeter to confirm wiring. If necessary, add a jumper so the two electromagnetic coils will be wired in series, as shown in Figure 7-14. Check to ensure that the sounder's arm moves freely and smoothly and that spring tensions are solid but not excessive. Next, slowly increase the DC voltage applied to the electromagnetic coils until the sounder's arm pulls in or "clicks." A selectable-voltage DC wall adapter is handy here, otherwise a group of regular AA or C cells can be used. Start with 1.5 volts and then carefully increase voltage up to an absolute maximum of 12 volts. Use a clip lead to quick-touch (make/break) one sounder connection. If the arm doesn't pull in with 12 volts applied, use your ohmmeter to check the terminals and wires for connection errors. After confirming proper sounder operation, hook up the whole works, as shown in Figure 7-14, send some Morse code with it, and you'll see that these critters are a barrel of fun!

The sounder's mechanism can be cleaned using emery cloth or steel wool, depending on its condition. Use your own discretion here. I acquired a sounder several years ago that had been scraped shiny clean with a pocket knife; it had deep scrape marks in it, but those smoothed out nicely with steel wool. After cleaning, polishing it with an appropriate product will really bring out the beauty of the sounder. If the base is in good condition, buffing with steel wool and then polishing with Old English or a similar scratch remover polish may make it look like new. If the base is heavily deteriorated, however, you may prefer to replace it. Check craft stores for wall plaque-type mounts that could be used as an attractive base for the sounder. If using such new wood, stain it, let it dry, and add gold leaf pin-striping for a beautiful appearance. Good looks are everything! You might even enlist the services of a local trophy shop to make a special nameplate or tag for your fancy base. Finally, team up the sounder with a pump key or bug of your choice and give it a prominent display position in your ham shack or den. Since current is used only when the key is closed (for instance, during your demonstrations), the battery life will be quite long. Now enjoy owning and using a genuine classic of yesteryear!

If you really want to have a ball with your new sounder, consider adding the optional receiver interface shown in Figure 7-15. The interface circuit plugs into your receiver's earphone socket and connects to the sounder's coils. You can then copy Morse off the air in true old-time style. You simply have to hear this setup in action to appreciate it: the incredible kalick-kalunk reports are marvelous! It won't sound *exactly* as it did in years past, however, because ham communications use International Morse whereas land line telegraph used Continental Morse, which was slightly more readable on sounders. Ah...but when you get good enough to actually copy half of a ham QSO on a sounder, you can really be proud of the accomplishment!

Refer again to Figure 7-15 and we'll take a look at how the sounder adapter circuit works. Incoming Morse code tones are stepped up from low to high impedance, rectified by the diode bridge, and the resultant DC voltage turns the transistor on/off in accordance with the Morse. When the transistor conducts, current flows from the battery to the emitter and on to the collector of the transistor, through the interconnected Morse sounder, and back to the battery. The

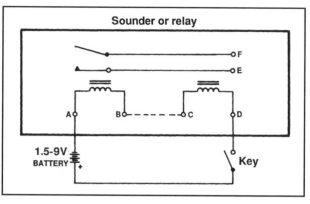

Figure 7-14. Wiring arrangement of dual electromagnetics used in most telegraph sounders and/or relays. Connections B and C are usually wired in series with a jumper (proper wiring is confirmed when connecting your ohmmeter between A and D gives a low resistance reading). Connections E and F are the relay contacts.

current flowing through the sounder's coils causes the armature to move in/out and produce "kalick" and "kalunk" sounds. I should also point out that sounders in the 30- to 400-ohm category work best in this circuit, and 4-ohm (large style) sounders should be avoided. Selection of components and circuit layout of this sounder converter aren't critical, as only audio frequencies and DC voltages are involved.

Transformer T1 is a small low- to high-impedance device you can get at RadioShack, but any "generic" substitute should also work fine. Diodes D1 through D4 are low voltage silicon devices. R1 is the bias resistor for the transistor. If you want, you can eliminate the on/off switch and simply connect the battery for operation. Also, if you've built this circuit before acquiring a sounder, a basic 25-ohm, 60-mA relay can be substituted to get you started; later you can replace the relay with a genuine sounder. If you want to hear actual American Morse, incidentally, tune in the net on 3544 kHz at 0030 or 1300 GMT daily. This group of sharp operators actually uses sounders with the converter previously described connected to their rigs.

Whichever path you pursue (using the sounder to copy off-the-air Morse or using it for a home display), I'm sure you'll find this project a sheer delight.

Build This Classic Transmitter from Amateur Radio's Golden Past

Our final project goes beyond mere "mild-mannered" homebrewing—get ready to build a magnificent reproduction of amateur radio's most famous old-time transmitter: the classic open-air Hartley (see Figure 7-16). This spectacular rig was "king of the hill" during the 1920s and '30s, and, for a ham, owning a replica of the little gem today is more exciting than owning a Model A auto, a Steerman Biplane, or other one-of-a-kind collectible. Indeed, this is the transmitter responsible for exposing newcomers to shortwave communication for almost two decades. It was made in both large and small versions, the latter of which actually worked Australia from the United States during the 1920s with only $1/2$-watt input! Needless to say, such a collector's item really adds a touch of glamour to any modern hamshack. Now you, too, can own an authentic piece of history that can only increase in value. Get ready for a real blast as we return to those exciting days of yesteryear when ham rigs looked like miniature moonshine stills and glowed with vacuum tube beauty!

Before diving into the assembly details, I must emphasize that this classic transmitter is primarily a fun project for showing off in your home rather than for actual on-the-air operation. Although it's quite capable of terrific on-the-air operation, its has some limitations as I've learned over the years. For instance, the transmitter doesn't "know" ham band edges and tunes a wide range of frequencies. Additionally, it is rich in harmonic output and requires using a modern antenna tuner or single-band antenna. For these reasons, I haven't included any type of coupling or connections for use with an outdoor antenna. Just build the little marvel as shown, tune in its signal on your general coverage shortwave receiver, and enjoy showing others a real amateur radio masterpiece!

The result of this project should be a perfect replica of a 60-year-old ham rig, so close attention to the minutest of details is extremely important. This involves using authentic-era parts throughout, which means about 80 percent of your time will be devoted to "scavenger hunting" components and only 20 percent of your time will go into final assembly. Good "hunting grounds" for parts include basements of old-time amateurs, remote corners of hamfests, and want-ad sections of various amateur magazines. Leave no stone unturned in your search—check out the oldest radio repair shops in your city, dealers carrying old radios, and even rural hardware stores. The components are still out there, but they aren't getting any more plentiful. Also, be aware that a "genuine article" usually commands a respectable price (like $20 to $30 for a No. 199 or 201 tube, $3 to $4 each for Acorn condensers, and $10 to $15 for a variable tuning condenser, etc.). Herein lies the true difference between "old junk" and "golden classics"! If possible, ask a local old-time amateur to assist you in evaluating the authenticity and quality of early radio components you find.

Additional parts you'll need to assemble the transmitter include a handful of brass screws and Fahnestock clips, No. 12 or 14 gauge bare solid wire, a four-pin base-mount tube socket, and a beautiful 1920 or 1930-style tuning knob. Here's a quick shopping list of components for your convenience.

1	.001-mfd Acorn Condenser
1	.000250-mfd Acorn Condenser
1	25-k ohm, 1-watt resistor
1	.01-mfd Acorn or paper-type condenser
1	2.5-mH RF choke
1	Variable tuning condenser (400 mmfd for 80 meters, or 250 mmfd for 40 meters)
1	Breadboard/base mount four-pin tube socket
6	Copper or brass Fahnestock clips, 24 brass screws
1	199 or 201A or 245 vacuum tube
10 ft.	(Approx.) $1/4$-inch copper tubing
1	small alligator clip to fit on $1/4$-inch copper tubing
1	old-time radio dial
1	authentic breadboard or awards plaque (obtained from a local craft store)

Plus: wood stain or shellac, small rubber feet, gold leaf pin-striping, hook-up wire

To ensure your success with this project, here are some helpful tips to keep in mind. Make the plate coil by slowly winding $1/4$-inch copper tubing on a $2\text{-}1/2$-inch-diameter form, such as a Comet cleanser can, a scrap piece of large diameter pvc tubing, etc. Wind 12 turns for 80 meters or 8 turns for 40 meters. Stretch the 12-turn coil to a length of $4\text{-}1/2$ inches and mate it with a 400-mmfd tuning condenser for 80 meters, or stretch the eight-turn coil to $4\text{-}1/2$ inches and mate it with a 200-mmfd tuning condenser for 40 meters. Although

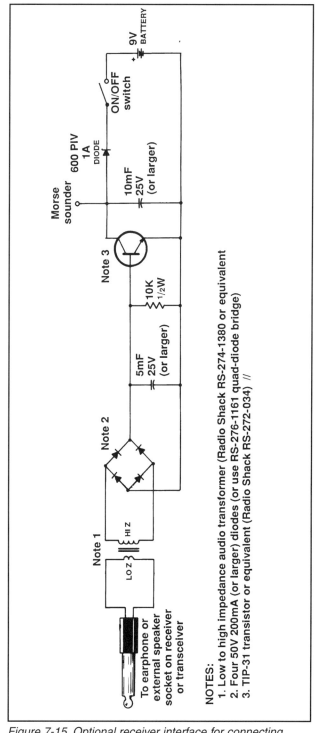

Figure 7-15. Optional receiver interface for connecting classic telegraph sounder to earphone/speaker socket of modern receiver or transceiver.

137

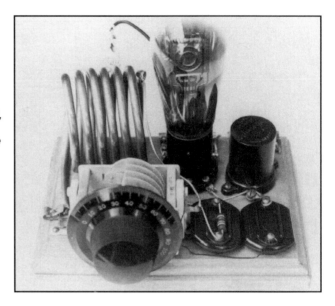

Figure 7-16. A reproduction of the famous Hartley transmitter. Note open-air construction, copper tubing coil, dome-envelope tube, and genuine National tuning dial. A beautiful classic!

this classic transmitter will operate on higher bands, frequency stability becomes quite critical and staying within ham band frequencies is challenging. After winding the coil, cut off excess tubing with a hacksaw, then flatten $1/2$ inch of the ends and drill holes in the flattened areas for mounting the coil with bolts/screws. This heavy-duty mounting approach is very important for coil rigidity and overall rig frequency stability, so avoid shortcuts. Finally, polish your copper tubing coil to a beautiful luster with steel wool.

Three types of classic tubes can be used interchangeably in this transmitter: the little peanut shaped UV-199, the dome-shaped UX- 201A, or the ever-popular UX-245 (also known simply as a 45-tube). Base pin connections for each of these three tubes is included in our schematic diagram's inset (Figure 7-17). The UV-199 and UX- 201A work fine with 90 volts on their plates. The 45-tube requires approximately 180 volts on its plate. Use batteries for powering this rig, not an AC supply (that will destroy the rig's beautiful tone!). Furthermore, using batteries results in a completely portable rig. The easiest way to procure the 90-volt power source is to snap together in series 10 common 9-volt batteries. Likewise, a string of 20 9-volt batteries connected in series will give you 180 volts of power. Purchase these batteries all at once and your overall cost will drop, certainly to less than the price of a classic 90-volt battery of yesteryear!

Filament voltage for the UV-199 and UX-201A tubes is 5 volts, which can be obtained in one of two ways. First, three 1.5-volt dry cells can be connected in series for 4.5 volts, which should prove quite soothing to the old-time tube's filament. Second, a 50- or 100-ohm 10-watt "slider type" adjustable resistor can be wired in series with a 6-volt lantern battery to obtain precisely 5 volts for the tube's filament. That same 50- to 100-ohm adjustable resistor approach will also work if you use a 45-tube, which has a 2.5-volt filament. Simply connect two 1.5-volt dry cell batteries in series, add the adjustable resistor (preset to its maximum resistance range), connect it to the transmitter, then slowly adjust the resistor for 2.4 volts at the tube's filament pins. Using batteries with this classic transmitter, incidentally, just adds to the authenticity! These rigs were popular before AC power lines graced our nation's landscape and batteries were their sole source of power.

Figure 7-17. Circuit diagram of the Hartley transmitter. Design is simple, but the rig works like a champ!

Since you're striving for 100 percent authenticity in assembling this classic transmitter, a rare, but still available, "body-end-dot" resistor should be sought. Color coding on these old resistors was marked as follows: body color = 1st number, end color = 2nd num-

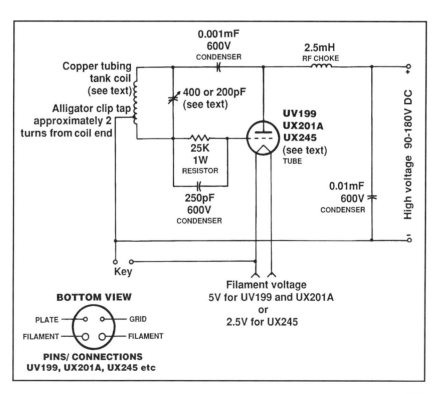

139

Figure 7-18. Top view of the Hartley transmitter showing layout of parts. Truly, this is the Mona Lisa of amateur radio...and you can build your own "original version"!

ber, and dot in middle of body = multiplier band. So a 25-k ohm resistor, for instance, will have a red body with green end and orange middle dot. A modern-day substitution (like that shown in my homebrewed replica) is acceptable until you can find the "real thing." Serious builders might consider brush-painting any body-end-dot resistor they can find with the proper color code, then drilling out its innards, and inserting a modern resistor (with wire leads connected to the original body-end-dot resistor's leads).

Acorn-type condensers are also rare, but quite worth the hunt! If necessary, large modern-style "domino"-type condensers can be used until you can find Acorns. Early Acorn condensers often had their values marked in mfd rather than mmfd (mmfd was an early designation for what is presently known as pfd), but you can easily understand/convert mfd to mmfd using the old "six digit counting technique." Basically this works as follows: a .000250-mfd capacitor equals 250 mmfd; a .001-condenser equals 1000 mmfd; 500 mmfd equals .0005 mfd; and 150 mmfd equals .000150 mmfd; etc.— just remember to count six decimal places.

The RF choke used in this transmitter is a standard 2.5-mH item; however, you could be really fortunate and find a black-encapsulated Hammarlund choke which can be screwed down to the wood board. Or you might uncover a tall, pink cloth-wrapped RF choke which looks even better than the one I used! Good luck hunting for these magnificent antique parts!

After acquiring all the parts, assembly of your classic transmitter begins by laying out the items on a wood board or awards plaque that's stained, polished, and approximately 8 by 12 inches (see Figures 7-16 and 7-18). Mount the tube socket with its large (filament) pin sockets toward the board's rear, position the copper tubing coil beside it, and firmly secure the coil and tuning condenser in place with brass screws and/or nuts. Remember rigidity is very important in the frequency-tuning area. Next, mount six Fahnestock clips along the board's rear, screw down the Acorn condensers, and mount the RF choke. Finally, interconnect the parts of the transmitter as shown in the schematic diagram of Figure 7-17. If you were fortunate enough to find a tuning condenser and tube socket with screw terminals, wiring may be accomplished using only a screwdriver or Boy Scout knife—a no-solder approach that's really 1930s-authentic! Sharp-eyed amateurs studying Figures 7-16 and 7-18 will notice only six coil turns and a small tuning condenser. I use my classic rig on 30 meters and it works great, but I incorporate many special tricks to ensure stable operation (definitely not recommended for newcomers!).

Check-out and operation of the assembled classic transmitter is quite simple, but I must emphasize: KEEP YOUR HANDS CLEAR OF THE CIRCUIT WHEN HIGH VOLTAGE IS APPLIED! Before connecting power, use your voltohmmeter to double check various circuit connections for continuity. You should read a perfect connection (0 ohms) between the high-voltage Fahnestock clip and the tube's plate socket pin, for example, and 25 k ohms between the tube socket's grid pin and the .001-mfd capacitor. Next, apply filament voltage (slowly!—remember this old tube has been asleep for many years and needs to be "awakened" carefully) and wait a few minutes. Next, connect the high voltage and tap a few brief dots on the key. You might even use a wooden pencil with sharp lead point to draw a small arc from the tank coil (wow, that is flashy!).

Next, try hunting for the transmitter's signal on a general coverage receiver. An old-time receiver works best here, as you can zip across a wide frequency range with a simple spin of the knob. If there isn't an antenna connected to that receiver, the only signal you'll hear will be the nearby transmitter. You can then decide whether to stretch coil turns, squeeze coil turns, or possibly short out a coil turn to change frequency coverage and "hit" your desired band of operation. Usually, shorting out a coil turn will raise the frequency. If you can't spot the rig's signal, remove the high voltage and move the coil's tap approximately one-half turn in either direction. Experiment with that tap until the transmitter is working smoothly and producing a clean signal.

Here a few more troubleshooting tips just in case you run into problems. Set your voltohmmeter to read current and connect it across the key. The transmitter should draw between 20 and 50 mA when operating properly. If no current is being drawn, recheck your wiring (which wire did you forget to include?). If current is above 100 mA, recheck condensers for shorts. But the best tip I can give you for tuning and using this classic transmitter is to find an old-time amateur who used such rigs in the Golden Age of Radio. These fellows are usually more than willing to help newcomers and their tales of operating with these classic rigs are incredible. In other words, this project can open a fascinating door to amateur radio's history. Enjoy!

Conclusion

I hope you've enjoyed reading about the projects in this book and that you'll build many of them. Homebrewing is one of the most exciting aspects of amateur radio, and everyone can join the fun. You may build one project this year and another project a year or two later, or you may build half the projects in this book in only six months. Every person is unique—and that's what makes each of us special.

In closing, I look forward to talking with each of you on the air during the very brief times we have available to enjoy our great hobby. I usually operate 14.180 to 14.225 SSB on Sundays between 2230 and 2300 GMT, and frequent 10.100 to 10.110 kHz weeknights between 0130 and 0150 GMT. Talk about a tight living schedule! If you write me, please include a Self Addressed Stamped Envelope (SASE) and be patient for a reply (I often work 16 hours a day, which leaves little time for letter answering).

Good luck, good homebrewing, and may the force of terrific radio signals always be with you!

73, Dave, K4TWJ

APPENDIX 1

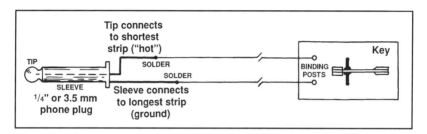

APPENDIX 2

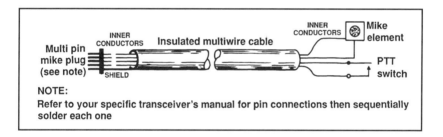

NOTE:
Refer to your specific transceiver's manual for pin connections then sequentially solder each one

APPENDIX 3

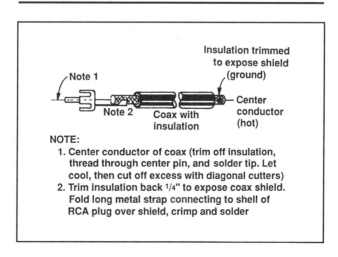

NOTE:
1. Center conductor of coax (trim off insulation, thread through center pin, and solder tip. Let cool, then cut off excess with diagonal cutters)
2. Trim insulation back 1/4" to expose coax shield. Fold long metal strap connecting to shell of RCA plug over shield, crimp and solder

APPENDIX 4

1. Trim end of coax cable with a sharp knife. Do not nick center conductor

2. Trim black jacket back 3/8" to expose braid. Cut braid end back 1/16" from dielectric to avoid electrical short

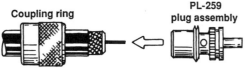

Coupling ring   PL-259 plug assembly

3. Slide coupling ring down cable, slide PL-259 on cable ensuring a smooth fit with center conductor. Screw PL-259 on coax, allowing inside threads to hold on black jacket

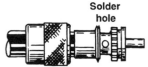

Solder hole

4. Solder coax center conductor to PL-259 middle pin, solder shield through holes in PL-259. A 250W soldering iron is suggested. Finish by sliding coupling ring onto body of PL-259

INDEX

alternator noise filter, heavy-duty 84
AMSAT 96
antenna wires and insulators 57
antennas,
    Bobtail, modified 68
    Bruce Array 74
    Carolina Windom 70
    Delta Loop 56
    El Toro 72
    invisible/secret 76
    mobile VHF/UHF, and mount 80
    mobile, high-performance 102
    phased (mobile) 110
    radio-active antennas 108
    turnstile (for satellite operation) 98
audio feedback 31

baluns 60
battery monitor 85
blinking eyes toys 2, 117
Bobtail Antenna, modified 68
Bruce Array 74

capacitors 14
Carolina Windom 70
charger, mobile 88
charger, triple rate 91
coax cable extension 27
coax cable, types of 58
coils 15
copper foil 32
CW keyer, remote-controlled 39

Delta Loop 56
dual IC transmitter 127

El Toro 72
electron flow 10
erecting antennas, easy method of 61

feedback, RF 30
field strength meter 35
filter capacitors 45
frequency-relation chart for RS-10/11 96
full break-in, simple concept for 51

glowing antenna 108
ground system, installing 32

Ham Radio Outlet 112
high performance mobile antenna 102

IC transmitter, dual 127
installation, mobile antenna/mount 80
insulators, antenna wires and 57
integrated circuits (ICs) 21
Invisible and Secret Antennas 76

Jan Crystals 127

I1

key adaptor 93
keyer
    kits, sources of 41
    remote-controlled CW 39

land line telegraph system 133

micronauts, pocket transmitters 125
mobile/mobiling,
    antenna, high-performance 102
    charger 88
    ground system 112
    phased antennas, for 110
    transceivers, spike/polarity protection for 83
    VHF/UHF antenna and mount 80

Modified Bobtail Antenna 68

NE-555 multivibrator 117
noise filter, heavy-duty alternator 84

Ohm's Law 10, 12
open air Hartley 135, 138

pen transmitter 125, 128
phantom antennas 31
Phased antennas 110
pocket transmitters (micronauts) 125
power strip 24

radio-active antennas 108
rapid charger 91
rejuvenating old transmitters 47
resistors 13
restoration of older-style shortwave receivers 43
robot operator, operation of 95
roaster, electric wiener 121
RS satellites 94

satellites,
    key adapter for working 93
    turnstile antennas for 98
secret antennas 76
shopping, hamfests 44
shortwave receivers, restoration of 43
skin effect 32
soil dampness tester 123
sounder adaptor 133
spike and polarity protection 83
SWRs 62

telegraph system, historical 129
Ten-Tec 41
The "El Toro" 72
The Carolina Windom 70
The Bruce Array 74
third hand pc board holder 3
toy, blinking eyes 2, 117
transmitters, rejuvenation of old 47
transformers 15
transistors 17
transmitter, build classic 134
triode tube 19
triple rate charger 91
Turnstile antennas 98

weiner roaster 121

zener diodes 21